The Cartographic Capital

Mapping Third Republic Paris, 1889-1934

Studies in Modern and Contemporary France 1

Studies in Modern and Contemporary France

Series Editors

Professor Gill Allwood, Nottingham Trent University
Professor Denis M. Provencher, University of Arizona
Professor Martin O'Shaughnessy, Nottingham Trent University

The Studies in Modern and Contemporary France book series is a new collaboration between the Association for the Study of Modern and Contemporary France (ASMCF) and Liverpool University Press (LUP). Submissions are encouraged focusing on French politics, history, society, media and culture. The series will serve as an important focus for all those whose engagement with France is not restricted to the more classically literary, and can be seen as a long-form companion to the Association's journal, *Modern and Contemporary France,* and to *Contemporary French Civilization*, published by Liverpool University Press.

The Cartographic Capital

Mapping Third Republic Paris, 1889-1934

KORY OLSON

Liverpool University Press

First published 2018 by
Liverpool University Press
4 Cambridge Street
Liverpool
L69 7ZU

British Library Cataloguing-in-Publication data
A British Library CIP record is available

ISBN 978-1-78694-096-4 cased

Typeset by Carnegie Book Production, Lancaster
Printed and bound in Poland by BooksFactory.co.uk

Contents

Figures

Acknowledgements

I have always loved maps. As I child, I often spent hours in front of my family's *Rand McNally US Road Atlas*. I would page through states and cities, follow the blue interstates over long distances, and wonder what life was like in the yellow cities, multi-coloured suburbs, or green parks of the many places I had yet to visit. In many respects, those days mark the start of this project.

It is impossible to thank everyone who deserves a mention. The first draft of this work was written at Penn State under the direction of Willa Silverman. Without her encouragement to combine cartography and French studies, I would never have got this far. I must also acknowledge Deryck Holdsworth who first introduced me to the key thinker and theorists in the History of Cartography and Denis Provencher who recognized my topic's importance and encouraged me to submit my work to Liverpool University Press. Of course, my graduate school cohort who cheered me on helped more than they know. Finally, a special thanks to the late Ruth Redhead, my first French professor, who told a 20-year-old young man at South Dakota State University that he could (and should) study French.

I must thank the School of Arts and Humanities at Stockton University. Deans Robert Gregg and Lisa Honaker supported and encouraged my project through many research grants, which allowed me to travel to French archives. Stockton University's Research and Professional Development Committee also recognized this project's potential on numerous occasions offering summer stipends and additional travel funds when I needed them the most. My fellowship with the Penn Humanities Forum gave me valuable feedback and access to the University of Pennsylvania's extensive resources at a time when it made a difference.

I would like to thank Alexandre Ragois at the Cité de l'architecture et du patrimoine, Archives d'architecture du XXe siècle, who helped secure many of the images for this book. Thanks, too, to the archivists at the Archives de Paris and Yale University Library for their assistance in finding maps and plans of Paris and to everyone at Liverpool University Press. Their substantive feedback improved my manuscript considerably.

I would also like to acknowledge friends and family for their support throughout this entire project. Lastly, I would like to thank my husband Lorin. His constant optimism and encouragement quickly became contagious and pushed me forward when I needed it the most.

Introduction

Creating Republican Paris

> Devant l'Hôtel de Ville, dans l'intervalle qui séparait l'ancienne Place du Châtelet de l'espace irrégulier qualifié Place de Grève, l'œil était affligé par d'horribles cloaques, nommés rue de la Tannerie, de la Veille-Tannerie, de la Vannerie, de la vieille Place aux Veaux, Saint-Jérôme, de la Vieille-Lanterne, de la Tuerie, des Teinturiers, etc. Cette dernière était si peu large, que la façade vermoulue d'une des maisons, en pans de bois hourdés de plâtre, qui la bordaient, essaya vainement de s'abattre: elle ne put que s'appuyer sur celle de la maison opposée. Et quelle population habitait là!
>
> —Baron Georges Eugène Haussmann,
> *Mémories*, Vol. III, 27–28

After Louis Napoleon's term as president and his 1851 coup d'état, Paris needed his attention. The city had undergone little significant renovation in the years since the 1789 Revolution. The narrow, unlit streets as described above by Baron Haussmann were a breeding ground for crime and disease. Bourgeois Parisians had grown weary of the uprisings that prematurely terminated the Bourbon restoration and the subsequent July Monarchy, bringing in the short-lived Second Republic. By mid-century, tired of crime and revolution, the Parisian bourgeoisie pinned their hopes on Napoleon III. His promises to restore order and become 'l'élu du commerce, de l'industrie, de la propriété' were just what they wanted to hear.[1] Many were tired of

1 In addition, Napoleon III played up stability while campaigning for the presidency of the Second Republic, proclaiming, 'que [le nom du président],

fearing not only for their wallets but their lives as well as they ventured into the night.[2]

The large concentration of unhappy working-class residents in the centre of the city worried both businessman and those in power. Yet beyond crime, early Second Empire Paris had other problems. Multiple cholera epidemics hit the city in 1832 and 1849. Although primarily concentrated in the dense urban core, this disease struck everyone and everywhere, alerting Parisians of the dangers of overcrowding. The first major outbreak killed 20,000 Parisians out of a population of 861,400 (Jordan 96). In addition to crime and disease, nineteenth-century Paris was known for having residents pour into the streets, erect barricades, and topple governments with authorities lacking the means to control them. Unable to suppress this population, leaders needed new methods for demonstrating their ability to rule over the urban space. Accordingly, one of the first ways the regime exerted its authority over the ever-mutinous city was through mapping. In 1853, under orders from the emperor, Haussmann's triangulation produced the first cadastral and topographic map of the city (Harvey, *Modernity* 112). In possession of this knowledge, Napoleon III and his *préfet* could begin to usher the troubled capital city towards modernity, as evidenced by new industry, scientific advancements, and social progress.

In addition to political gain, stability would prove necessary for economic development. France's violent revolution and the continental Napoleonic wars slowed the nation's industrial revolution; but it did arrive. As with eighteenth-century England, the creation of France's industrial-based economy required a large, inexpensive workforce. The impoverished rural peasant looking for work soon became the factory worker, whose cause was made famous by Marx and Engels. In France, this population flooded into Paris from poorer rural departments, overcrowded the city centre, and changed the city's demographics. With industrialization, the French bourgeoisie's wealth increased, but

accepté par tous, soie un premier gage d'oubli, de réconciliation, car sans la réconciliation des classes, plus de paix, plus d'industrie, plus de crédit, mais la misère et l'anarchie' (Anceau 129).

2 Crime was a major topic of discussion in nineteenth-century Paris. While 'the bourgeois felt secure behind their triple bars and bolts,' they eventually needed to traverse 'nooks and corners ideally suited to robbery with violence by day and night' (Chevalier 1–2).

an ever-present militant working class disrupted governments, made the city dangerous for its leaders, and gave the French capital a bad reputation both at home and abroad. If this was modernity, the French ruling class had to ask if it was worth the trouble. But the emperor had previously expressed interest in helping those less fortunate. In exile, Louis Napoleon, a believer in Saint-Simonism, felt that the government should assist its citizens. The Comte de Saint-Simon's ideals, such as the considerable emphasis placed on the moral virtues of industrialism in terms of its capacity to improve the standard of living of the poor,[3] fit in well with the emperor's plan to use modern technology to alleviate the misery of the working class. During his imprisonment at the prison of the château of Ham during the Restoration, the emperor published a pamphlet of his *Des idées napoléoniennes*, many of which included government programmes to improve society (Lentz 92). In his pamphlet, Napoleon III proclaimed that government was not 'un ulcère nécessaire, mais c'est plutôt le moteur bienfaisant de tout organisme social' (3). In addition to 'une magistrature considérée qui fasse respecter la loi,' a good and competent administration 'améliore le sort des classes pauvres, qu'elle éveille toutes les industries, et qu'elle tient une balance égale entre les riches et les pauvres, entre ceux qui travaillent et ceux qui font travailler, entre les dépositaires du pouvoir et les administrés (38–39). The emperor also privileged extensive public works, since they 'favorisent un grand progrès social. La création de nouvelles routes et de canaux augmente la valeur des terres et facilitent l'écoulement de tous les produits' (61). These socialist ideals forecast a nation and a capital city whose citizens would benefit from all that French industry and technology could offer.[4] The emperor believed that his version of government would improve society. Yet this path to prosperity did not last. Napoleon III's dismissal of his *préfet* in January 1870 was followed by defeat on the battlefield at Sedan that same year, which ended Bonapartist rule in France and the continuation of the Haussmannian dream for Paris.

3 In his *Nouveau christianisme*, Saint-Simon argued that these values corresponded to the ethics of true Christianity, and thus that no Christian could deny the validity of his doctrine (*Selected Writings* 27).

4 In *Paris, Capital of Modernity*, David Harvey highlights Louis Napoleon's time spent imprisoned at Ham prison. There, the future emperor published a pamphlet entitled *L'Extinction du paupérisme*, in which he championed the right to work as a basic principle and proposed state legislation to set up associations of workingmen (69).

Hoping to combine the Second Empire's push to modernize with democratic values, yet eager to distance itself from the rest of the turbulent nineteenth century, Third Republic administrations wanted to promote France as a unified nation with Paris, its capital, as the model example of stability. Taking into account the violent end of its predecessor, the republicans emphasized their differences with the Second Empire while reassuring business leaders and foreign governments that maintaining control in Paris was achievable without suppressing newly granted rights. They believed it plausible for the new regime to take the best of the imperial city and make it republican. To what extent did maps of Paris become instruments of power and how did their representations of authority change the multiple messages contained in maps? How was the past, primarily nineteenth-century improvement and embellishments to the city of Paris, along with the themes of modernity and security, portrayed in Third Republic cartography? How did the bourgeoisie and the successive governments it manned look to and plan for the future?

Although the terms 'bourgeois' and 'bourgeoisie' can be very broad, I employ them in this book to describe the developing Parisian middle and upper middle classes; the people who benefited most from Napoleon III's stability. For this definition, I turn to Philippe Perrot, whose work on the nineteenth-century French bourgeoisie suggests that industrialization, combined with the surge of consummation during this time, created a society that progressed 'sur la double voie de la prospérité matérielle et de l'égalité civile' (192). It was this demographic who would fill much of the nation's bureaucracy, survey the land, create its cartography, and read the published maps. I am also heavily inspired by David Harvey's 'bourgeois modernity,' defined as 'founded on the rock of private property and sought freedoms of speech and of action in the market, and the kind of liberty and equality that goes with money power' (85). This book seeks answers to these questions by examining how Paris is portrayed in the maps of the capital city published under the Third Republic, a time when the city underwent significant demographic growth as well as when technology and science changed how people used and understood the city. In addition, it seeks to understand how those maps promoted the post-Commune city as modern and, finally, how they portray the evolution of city space, along with the changing definitions of Paris itself.

Cartography as Print Culture Studies

The study of maps as instruments of the new regime provides additional layers to our understanding of the ideology and policies of the Third Republic. Within the study of history of reading, reading practices, and print culture, an analysis of map production and reception is useful for demonstrating what advances were being made at the juncture of cartography and the publishing industry and how those advances were received by map readers. For example, the use of colour in maps changed dramatically in the second half of the nineteenth century and became even more varied as the twentieth century progressed. Additionally, print culture studies provide insight into distribution methods and technologies available to accommodate map publishing. With the passage of the hallmark 1881 law instituting freedom of the press, map publishers had more liberty in how they portrayed the city and government, though many chose to shine a positive light on Paris as government favour meant a better chance at securing lucrative textbook contracts (Amalvi 191).

French publishing under the Third Republic made significant advances. Political and economic factors combined with technological developments to change the nature of book production. To explore the varied French perspectives on Paris, this study relies on approaches and methods in cultural history. As a combination of print culture and history, this project fits within the greater context of cultural history, drawing primarily on the formulations of Roger Chartier regarding the relationship between identity and power (see 'Le Monde comme représentation'). Though researchers have examined the relationship between map discourse and the notion of modernity in its effect on an urban population, maps of Third Republic Paris as an expression of modernity remain under-examined. Studies in this area address mostly French travel or tourism, and centre primarily on the texts supporting these maps, as in travel guides.

My research will add to several previous studies by Catherine Dunlop, J. B. Harley, Dennis Wood, Richard Dennis, David Harvey, Henri Lefebvre, James Lehning, and others. I build on Catherine Dunlop's investigation of mapping and counter-mapping around the new Franco-German divide following the nation's 1871 defeat and subsequent return of Alsace-Lorraine to French control in 1918. In *Cartophilia* (2015), Dunlop studies various approaches to mapping the international boundary, many of which arose from popular cartography. My book, on the other hand, focuses on greater Paris, the population, political, cultural, and commercial capital of the nation. Cartography commissioned by the government and drafted by official agencies or urban planners to accompany urban legislative proposals, such as the 1919 loi Cornudet, presents the government's view of the changing metropolitan area.

While he investigated topographical maps published by the Ordnance Survey in the nineteenth century, J. B. Harley's work primarily concerns England or Western Europe as a whole; he did not apply his theories to Paris. Building on Harley's examinations of discourse and control in *The Power of Maps* (1992), historical cartographer Dennis Wood studies maps as semiological systems. While maps may appear neutral to the reader, '[cartographic] system[s] can be corrupted' (105), Wood maintains, acknowledging the desire by those in power to portray geographical areas as mythic objects. Applying Roland Barthes's semiological schema, in which the French philosopher 'demythified' common myths in post-war France, to maps, Wood asserts that they do not escape the 'the grasp of myth' (105). A positive view of Paris, for example, would affirm to readers that republican reforms worked and benefited citizens.

Although he did not focus on maps himself, Barthes demythified images found in 1950s French advertisements. Through his schemas, he challenged the innocence of images and demonstrated that the connotation often involved an imposed meaning (181–233). Like many of Barthes's *Mythologies*, maps have become commonplace in twenty-first-century society. We have come to expect them in travel guides, phonebooks, newspapers, and advertisements. Equally important, the contemporary reader has faith in maps and regards them as natural, ordinary objects. '[A map] is mythic [due] to the degree that it succeeds in persuading us that it is a natural consequence of perceiving the world' (Wood 105). The reader sees the world through maps. What is not known to many of those readers is that the final cartographic presentation is provided by a partial cartographer.

How might this translate to the French capital? An 1841 law mandated that all rail lines terminate in Paris, so many travellers were forced to change trains there when crossing the country. If such a passenger consulted a guidebook map, Haussmann's new boulevards, usually the most prominent streets visible on maps, would 'guide' them from one large train station to another or, if needed, to the transformed historic centre of Paris. To de-clutter the document, cartographers often removed many side streets. Although this action would render the cityscape less confusing, it also becomes less detailed. The new simplified city plan took away distractions, but also choice. For an unseasoned visitor to the city trying to find his or her hotel, guidebook maps made it almost impossible to stray from Haussmann's wide boulevards into the city, regardless of whether that was the quickest or easiest path to their destination. Later, in the 1930s, Prost's red autoroutes stand out and assure readers that they could employ their personal automobile throughout the Greater Paris region with ease, even if the construction of those limited-access roads disrupted the lives of many working-class suburbanites. Wood's map analysis reminds readers that what they see may not represent the whole truth. He thus points out the dangers of simply accepting mapped representations without question and reinforces the need for a multidisciplinary approach to the study of cartography.

Adding another important layer to the map theories of Harley and Wood, Richard Dennis investigates nineteenth-century urban space. His studies of British suburbs present a model on the attractions of living beyond the city centre. Similar to the region surrounding London, the construction of the suburban railways created a market for single-family homes in *la région parisienne* during much of the twentieth century, though not on the same scale as seen in the United States or Britain. David Harvey provides an in-depth look at the discourse of modernity in Napoleon III's Second Empire in *Paris, Capital of Modernity* (2003). For Harvey, 'a fundamental question is how Paris worked and how it became such a central site for the birth of modern' (18). He focuses his studies on Second Empire France, at the onset of the Industrial Revolution. Harvey holds a capitalistic view of modernity, which he defines as corresponding to the period in which finance was used to transform both society and the city.[5] Modernity for Harvey can thus be viewed as the moment in which districts of Paris were transformed from working-class and integrated

5 Stating that 'the circulation of capital is in charge,' Harvey uses Balzac's descriptions to demonstrate how Paris, as Balzac saw it, was a city 'devoid of morals, principles and genuine feelings,' but how 'everything stimulates the upward march of money' (32–33).

neighbourhoods, where all classes commingled, to a city centre used for government, spectacle, and recreation (cafés, department stores, and boulevards), with the working classes pushed further out, to the periphery.

Although a prime destination for French and European travellers offering much to see and do, Paris also had many problems in the mid-nineteenth century. It was, in Haussmann's metaphor, a 'sick' city, and the emperor appointed the *préfet* in the early 1850s to cure Paris's ills. Even though significant progress had been made by 1861, there was more for 'surgeon' Haussmann to do. The *préfet* often used this terminology of sickness himself when referring to the capital:

> After a prolonged pathology, the drawn-out agony of the patient, the body of Paris, was to be delivered of its illnesses, its cancers, and epidemics once and for all by the total act of surgery. "Cutting" and 'piercing' were the adjectives used to describe the operation. Where the terrain was particularly obstructed, a 'disembowelling' had to be performed in order that arteries be reconstructed and flows reinstated. (Vidler, qtd. in Harvey 260)

If Haussmann's city was a patient, the cure was thus modernity.

The Cartographic Capital will continue Harvey's examination of Second Empire modernity into the Third Republic, when the desire to be seen as modern remained just as important to the new French government. Republican modernity emerges in multiple mapped forms. In cities, for example, it is manifested through the inclusion of technological achievements such as sanitation and transportation (railway lines figure prominently). In addition, education – free, secular, and now obligatory – appeared through the construction of schools in numerous neighbourhoods throughout the city. Boulevards, another important part of the urban landscape during the Second Empire embellishment of central Paris, remain prominent on maps though perhaps more as a former enhancement upon which administrations could accentuate their own modernity – for example, through naming – than as an example of republican action. After 1871, however, much of the remaining incomplete parts of Haussmann's grand plan were finished, albeit reluctantly. The Avenue de l'Impératrice, for one, lost its original imperial name. Following the arrival of the Third Republic, the majestic road between the Bois de Boulogne and the Arc de Triomphe first became Avenue Général Uhlrich, then

Avenue du Bois de Boulogne. Well after Haussmann's departure, the government happily appropriated his works as they built out the city.

Henri Lefebvre compliments Harvey's views on modernity, especially when viewed as an economic asset to the Third Republic. He affirms that '[social] space is a [social] product.' For him, space is a means of control, and hence of domination, of power (26). With the bourgeoisie appropriating much of central Paris as their own, the central *arrondissements* monopolized much of the city's maps. Jean-Charles Adolphe Alphand's atlas *Les Travaux de Paris, 1789–1889* demonstrates the extent to which the bourgeoisie controlled the cartographic conversation (see Chapter 3). Of course, the instrumentality of space was not limited to economic productivity, for it could also facilitate political engagement (Lefebvre 411). Paris dominated the nation's industrial capacity, but most factories occupied the urban fringe; a space all but ignored in 1889 by Alphand. However, the concentration of workers beyond the city limits produced the political victories that led to the *cartel des gauches* picking up parliamentary seats in 1924 and the Front Populaire's 1936 rise to power; shifts that forced cartographers to acknowledge this demographic on their documents. Although map-makers had preferred to silence this space in the past, its electoral power now eliminated that possibility. By 1934, Prost's *Carte générale* went as far as to silence Paris and used colour to display the growing suburban ring's new economic and political influence (see Chapter 6).

Moving from the political and economic, Eric Hazan's work on space explores the shifting ideas of boundaries and identity. His *L'Invention de Paris* (2002) looks at Paris's internal divisions, primarily as the city absorbed and annexed various villages and neighbourhoods as it grew throughout the nineteenth century. Although initially resistant to joining Paris, those *intra-muros* community residents would eventually benefit from the prestige that comes with living within the city limits. Since a person's residence or neighbourhood forms an integral part one's identity, locational definitions are both personal and political.

For the inhabitants inside the wall, for example, Hazan shows us how even established identifications can be cause for debate. The exact boundaries of Montmartre or the Latin Quarter, for example, may differ depending on whom you ask. In addition, a contemporary resident of Belleville may claim its burgeoning diverse and artistic character as their own, but that same address a hundred years ago would have elicited a very different reaction from a fellow Parisian. Politically and financially, differences over boundaries can be far more serious than a common stereotype (Hazan 5). Today, the role of border and identity, for example, can determine tax or insurance rates or whether your child attends a certain school. Official divisions, useful to the tax man and the police, are often endorsed by cartographers.

Hazan's guided tour of the city describes the historical urban space, but also explains the consequences of Paris's growth. In cities such as Los Angeles, Tokyo, or Berlin, borders are an ambiguous zone of constant movement. Paris, on the other hand, has been constrained by its city walls throughout its history. Its growth pattern has always given it a more or less circular form, and it has only been able to extend in a succession of circular rings (Hazan 6). Multiple physical constraints also played a psychological role in Parisian identity. The scenario never changes. A new wall is constructed, with enough room to provide much-needed free space, which quickly gets developed. Land within the wall becomes scarce, buildings are pressed together, plots fill up. Outside the wall, and despite the laws against it, new houses appear (Hazan 8–9). And the cycle repeats. Adolphe Thiers's final barrier establishes the ultimate Paris city boundary. However, throughout the nineteenth and twentieth centuries, maps show us to what extent this growth pattern continues. The topographical Paris basin limited the urban sprawl somewhat, but the notions of the city and its space in Greater Paris expanded as more newcomers arrived each year.

Social, political, and economic divisions went beyond the Paris-suburb border. With much mental and financial attention devoted to improving Paris throughout the Second Empire and the Third Republic, resentment surfaced in the rest of France (Harvey 317). This contentious relationship between Paris and the provinces has never fully disappeared.[6] However, the need for republican leaders in Paris to homogenize the nation state remained. By sending out loyal *préfets* and *instituteurs*, the government not only increased its presence in the countryside but also facilitated the spread of republican ideals

6 For an in-depth look at the relationship between Paris and the provinces, see Bernard Marchand, *Les Ennemis de Paris: la haine de la grande ville des Lumières à nos jours*.

and control in Paris (Hobsbawm, *Tradition* 105). Regardless of the nation's feelings towards its capital, Paris was an excellent showpiece and France used it as such. The city hosted large-scale public events, such as World's Fairs, prominent funerals, and the state celebrations of Bastille Day. When held successfully, these events reaffirmed the regime's control over the once-chaotic city.

Echoing Lehning's and Weber's theses through cartographic discourse, published maps represented and endorsed the republican regime and were used to publicize a stable, modern capital city. The fact that the French (male) voting population would now play a more active role in a representative democracy than under the increasingly authoritative empire made it an ideal target audience for these messages as government officials developed 'an interest in promotion and dissemination of national values through schooling and military education, as well as through the invention of public national ceremonies, rituals and symbols' (Hobsbawm, *Nations* 107). Cartography could be merged into these values, at which time they began to endorse the new regime, its ideology, and its benefits to France both as a democratic republic and as a member of a changing European society.

Throughout the nineteenth century, Europe transitioned from a continent of kingdoms and empires to one composed of nation states. While Hobsbawm cites language as 'the decisive or even the only criterion of potential nationhood' (*Nations* 102), the shift was also ideological and political: 'the standardization of administration and law within it, and in particular, state education, transformed people into citizens of a specific country' (Hobsbawm, *Invention* 264). The European city was becoming more democratic and maps readily portrayed this shift. In Paris, after years of paying for Haussmann's transformation of the city centre, citizens from outlying districts were ready for the outer, most recently annexed *arrondissements* to be improved.[7] As the administrations reacted and pushed boulevards,

7 Haussmann had largely ignored the outer ring of the city, annexed in 1860, despite its expanding industries and population. He had good reason for concentrating on the centre, and with a sympathetic, nominated City Council he could ignore the demands of the outer areas. But under the Third Republic it was no longer possible to do so. The representatives of the outer *arrondissements* in the new City Council combined their influence to obtain a fair share of public investment for their own wards (Sutcliffe, *Autumn* 46).

parks, and sewer and water lines further out, the government mapped the substantial undertaking, further documenting the large investment the Republic had made all across Paris.[8]

The Third Republic's attention to Paris's outer *arrondissements* epitomized the desire to unite the city. For France, a traumatic defeat on the battlefield, the loss of territory, and the establishment of the Republic led to not only the formation of a new government, but also of a new ideology. A uniform Paris could serve as an example for the entire nation in a time of need. In *Paris Between Empires* (2001), Philip Mansel suggests that a decisive step in the 'nationalization' of Europe took place in 1870, with the defeat of the French by Prussia (429). The proclamation of the German Empire at Versailles disrupted previously established alliances and led to the eventual formation of new nations in Central and Eastern Europe. This event also triggered nationalist sentiments in France. Eager to find reasons for the French defeat, the national government looked at internal deficiencies. One solution was to examine France's multiple internal cultural and linguistic differences. National leaders decided that a more homogenous nation was needed, and successive republican governments turned their attention to the promotion of a single unified French identity. Maps, through this discourse, show the end result: a modern, secure, republican capital for a strong united France.

Overview

Following this Introduction, the chapters in this book are arranged both chronologically and thematically. By documenting the transformation of Paris throughout the Third Republic through maps, I am able to identify related trends, themes, and changing ideologies.

Chapter 1: 'Working with Maps.' The cartographic investigations in this book focus on both the challenges and insights to be gained from working with maps. To understand historical maps fully, one must identify a past map-maker's cartographic discourse, the interaction between map-maker and map reader. This interpretation can be

8 Adolphe Alphand's atlas *Travaux de Paris, 1789–1889* (1889), discussed in Chapter 3, is a prime example of how the city showed the growth of infrastructure improvements such as water, sewer, and roads.

complex and each individual reads a map according to his or her particular background. In order to understand fully the proposed communication circuit, one may turn to a variety of tools, such as semiotics, the framework for my map image analysis. The investigation of colour, shapes, symbols, and text on maps of Third Republic Paris help uncover underlying themes of modernity, stability, ease of movement, and growth. There are also benefits to be gained from working with maps. Whereas legislation and newspaper articles provide precise detail regarding historical events, maps provide a more global view of a past time or space. The visual nature of the medium has the potential to draw a reader's eye much more effectively than pages and pages of black-and-white script.

Beyond discourse, this first chapter also investigates the changing role of the French state in the history of cartography. By the nineteenth century, the government improved access to maps, via education reforms that included the study of geography. With a population that could more readily access and understand maps, cartography helped foster the growing field of French urbanism and planning. Furthermore, the government shifted from presenting what it had accomplished in Paris throughout the Third Republic to planning and managing its growth, and state cartography needed to adapt. An investigation of historic cartographic colour printing techniques will show how this was done and support this book's map analysis.

Chapter 2: 'Creating Map Readers: The Rise of Geography and Cartography in Nineteenth-Century France.' My initial chapter examines the creation of French map readers during the French Third Republic. Throughout much of history, cartographic literacy had been the preserve of a small, educated elite. But the loss to the cartographically proficient Prussian army in 1870 initiated a drive to introduce geography and maps into French society. One way to do that was through education. Jules Ferry's 1881 reforms made primary school *laïque* (secular), free, and obligatory. Over time, these dictates created a better-educated citizenry. Vidal de la Blache and Jules Simon built on Ferry's reforms. Their administrative work laid the foundation for the large-scale introduction of geography into schools. Once in classrooms, textbooks such as Fouillée's *Le Tour de la France par deux enfants* taught the benefits of geography to the French. Most striking: Fouillée included regional maps in her text in 1884. This cartographic introduction provided immediate millions with access to small, simple maps for the first time, effectively democratizing them on a large scale. Although Fouillée's narrative utilizes a republican vernacular that relied on the natural and geographical names of the departments which her protagonists visit, she employed the much more familiar, but outdated, provincial names to label her cartography, making it easier to understand.

As Fouillée introduced maps to schoolchildren, Vidal de la Blache promoted physical geography in France, the study of which would improve the topographical accuracy of future maps. Furthermore, similar to Fouillée's desire to update an outdated provincial vernacular with a newer republican departmental one, Vidal too changed how the French understood and viewed their national territory. He followed a European shift away from reliance on historical jurisdictions towards new physical ones, such as those based on river valleys. Yet perhaps more important than a re-examination of its own terrain, his work proliferated geographic training in France. He authored textbooks that helped create qualified geography teachers, allowing students to receive competent instruction in the discipline. In addition to the above educational and science texts, I examine how French cartographic literacy grew in other ways, such as rail and colonial expansion under the Third Republic that brought more maps into circulation. Finally, as the French learned how to read cartographic documents as the Third Republic progressed, I look at the changing nature of map publishing at the end of the nineteenth century and how that transformation affected future cartographic products. Enhancements in technology facilitated a more robust cartographic publishing industry, one in which printing in colour became easier and less expensive and, as we see throughout the book, more prominent.

Chapter 3: 'The Triumphant Republic: "Paris en 1889: les opérations de voiries exécutées entre 1871 et 1889."' The year 1889 was arguably the first important milestone celebrated by the Third Republic. A century after the fall of the Bastille and almost 20 years into the new regime, this chapter examines Jean-Charles Adolphe Alphand's *Travaux de Paris*, an atlas containing a series of maps that present the spatial and technical organization and amelioration of the city from 1789 to 1889 and published for the upcoming World's Fair that same year. Through the 16 *planches* (map plates), Alphand emphasized the continuity that he felt characterized the various transformations of the French capital throughout the Third Republic. I focus specifically on *planche* 13, 'Paris en 1889: les opérations de voiries exécutées entre 1871 et 1889,' as it documents the government's ambitious Parisian road-building drive, in many ways a continuation of Haussmann's Second Empire campaign, but now with republican support. Alphand's 'Opérations de voiries' employs bright yellows and reds to celebrate the city and reaffirms the importance of an effective road network to Paris. By planning and building these streets and avenues, and then publishing a map documenting its progress, the government proved its commitment, at least on paper, to bettering all of the city's neighbourhoods. With a surplus of yellow development in the outer *arrondissements*, Alphand subtly reaffirms that the Republic has concentrated its time and energy out of the bourgeois centre and in neighbourhoods previously ignored by Haussmann.

In addition to demonstrating a more egalitarian approach to Parisian betterment, Chapter 3 argues Alphand's work here also conveys authority and stability in the early years of the Republic. Alphand's 'Opérations des voiries' map reminds its citizens that, via the Service des Travaux's road works, an acquisition of knowledge of the entire city – via significant surveying and demolition – would not only improve the movement of people and goods but also translate into enhanced jurisdiction over once-rebellious or squalid neighbourhoods from which too many nineteenth-century revolutions and revolts had erupted. Perhaps more important is its presentation of the city. On this *planche* 13, the government's definition of 'Paris' ends at the wall built by Thiers in the 1840s. Alphand's bright colours draw the reader's attention to the significant infrastructural improvements, but his energies and attention remain faithfully within the 1860 city boundary. By limiting his time and efforts to this space, Alphand reinforces a bourgeois Paris-centric view of the capital, and suggests that any existing development beyond that line does not merit cartographic representation in a map representing Paris.

Chapter 4: '1906: A New Way of Seeing Paris: The Service Géographique's *Carte de France*.' Alphand's *Travaux de Paris, 1789–1889* demonstrated his knowledge of the capital city. By the turn of the twentieth century, a greater familiarity with physical geography, thanks in part to the efforts of Vidal de la Blache, provided the Republic with an educated workforce able to document its entire territory more professionally. I examine the century-long evolution of the Dépôt de la Guerre's *Carte d'état-major* and the subsequent Service géographique's

Carte de France. I focus more specifically on the army's 1906 'Paris' sheet (*feuille* XXIII_14) of the *Carte de France*, which documented the capital city within the greater Seine department. Complementing its portrayal of the French capital, I also investigate the Paris map sheet to demonstrate how the government used the Dépôt de la Guerre and its successor the Service géographique to professionalize and homogenize its state cartography at the start of the twentieth century. I look at the history of the French government's surveying and mapping of the national terrain throughout the nineteenth century. Napoleon first proposed updating the cherished Cassini map series when he felt that a comprehensive cadastre map would complement his Code Civil. Military defeat ended any grand Napoleonic surveying force, but subsequent regimes employed the Dépôt de la Guerre to survey and map most of France. By the turn of the century, the recent 1871 loss to the Germans reinforced the idea that military cartographic reform was needed. As the Dépôt de la Guerre became the Service géographique de l'armée, its maps themselves showed age. Some of the army's surveying predated the 1827 inaugural Saint-Etienne-Andrézieux rail line. In order to properly defend the country, the army needed its maps to include the nation's current infrastructure, industrial development, and urban growth, and it decided to abandon the outdated *état-major* map for the new *Carte de France*. The Service géographique needed not only to portray the nation accurately, but to do so in a manner that most citizens could understand. So even simple choices such as font and colour had significant impact on how people viewed and understood the national terrain. With its more professional and sizeable corps of *ingénieurs géographes* (geographic engineers), the government had more data in its possession and thus liberty in how it presented the nation.

This Paris sheet was not only one of the best recognized sheets from the series, it also presented the army's, and by default the government's, comprehensive view of the growing capital city, which now extends well behind Thiers's Wall. *Feuille* XXIII_14, one of the first nine *Carte de France* map sheets to be published and made available to the public, served as part of the larger introduction to the nation of the new modified *état-major*. The Service géographique hoped that *feuille* XXIII_14 would be more user-friendly and commercially viable than the previous editions with their smaller scale and outdated information. By the turn of the twentieth century, new suburban industrial growth has pushed the outer edges of the urban area well past Thiers's wall.

No longer portrayed by Alphand's 1889 'Opérations des voiries' as an isolated bourgeois 'island,' Greater Paris had now breached the wall, filled the Seine department, and overflowed into surrounding departments.

Chapter 5: '1919: The Beginning of French Urbanism: Léon Jaussely's *Plan d'éxtension*.' The Service géographique de l'armée's *Carte de France* accurately documented the nation's physical terrain. However, it also served as a base for two future maps (discussed in Chapters 5 and 6). This chapter examines the first of these: Leon Jaussely's 1919 *Projet lauréat de la section générale du concours du plan d'extension de Paris*, produced as a result of the 1919 loi Cornudet. In 1915 François-Joseph Cornudet des Chaumettes, a deputy from the Seine-et-Oise introduced to the Assembly the first piece of legislation that would require the creation of a comprehensive regional plan for Paris and its environs. The law proposed that any French city with a population greater than 10,000 would need to have a 'projet d'aménagement, d'embellissement et d'extension.' Why was such a law needed? The growth of Paris throughout the Third Republic, aggravated in part by the introduction of the personal automobile, pushed residential, commercial, and industrial development well beyond Thiers's 1840s walled fortifications, which also served as the city's official boundary. The legislation also grew in part out of the influence of the Musée social movement. Founded originally as a forum for research based around documents from the 1889 Exposition Universelle's social economy section, one of ten official groups that constituted the large-scale event, the Musée social became the focal point of studies of hygiene, social reform, and ultimately urbanism. It grew in popularity and influence in part due to its role in the debate over the dismantling of Thiers's fortifications. Under the leadership of men such as Eugène Hénard and Léon Jaussely, the Musée social pushed for improved regulation of growth and the incorporation of more green space into French urban agglomerations. Jaussely joined with Roger-Henri Expert and Henri Sellier to submit their *Projet lauréat de la section générale du concours du plan d'extension de Paris* to a jury of politicians, bureaucrats, and academics. Their entry won and took home the Première Prime prize in the Section I category, reserved for those projects that incorporated maps and a regional plan as defined by the loi Cornudet.

Chapter 5 argues Jaussely used the *Plan d'extension* to address what he, Expert, and Sellier wished to see in a modern French metropolis.

Contrary to Alphand's 'Opérations des voiries' and the army's *Carte de France*, both of which documented the past, Jaussely's sizeable hand-painted *Plan d'extension* shows the future and marks the beginning of modern urban planning in France. Influenced by the French school of urbanism and the Musée social, Jaussely incorporated their recommendations for cities to provide healthy living environments for residents and visitors. New parks and *cités jardins* in the suburbs incorporated usable green space in all corners of the area. New ports, *aérogares*, and railway stations on the agglomeration's edge ensured ease of movement over the large and growing expanse of territory.

Chapter 6: '1934: The Rise of Suburban Paris: Henri Prost's *Carte Générale*.' With urbanism an accepted practice in France by the 1930s, this chapter examines the 1934 *Carte générale de l'aménagement de la Région parisienne*, a brightly coloured, multipage representation of Paris and its suburbs. The city's growth outpaced the planning dictates established and eventually abandoned by the 1919 loi Cornudet. By the early 1930s Prime Minister Poincaré recognizing the need for new and larger roads, pushed for a new planning regime for the capital's region. As a result, the National Assembly passed broad planning legislation and brought the Paris region and its planning under the control of national government. Parliament passed 'la loi du 14 mars 1932,' which for the first time officially defined *la région parisienne* geographically, not demographically, as the area within a 35-kilometre radius from the *parvis* of Notre-Dame. In the legislation, the government formally created the Comité supérieur d'aménagement de la région parisienne (CSARP), previously known as the Comité supérieur de l'aménagement et de l'organisation générale de la région parisienne (CSAORP). Charged with drafting a comprehensive plan and presided by Louis Dausset, the 40-member commission looked over and chose Henri Prost's *Carte générale* and named him Urbaniste en chef du Comité Supérieur. The law also required a *plan directeur* for the region and on 14 May 1932 Prost provided his *Carte générale*: a document that introduced the suburbs as well as a plan for the region's future.

Prost's map, the last officially approved cartographic proposal for the capital published under the Third Republic, created an urban plan for the Greater Paris region. It recognized the changing nature of early twentieth-century cities, where the automobile both enhanced personal movement and overwhelmed nineteenth-century infrastructure. The CSARP, Paris's regional planning authority, had asked potential planning candidates, including Prost, to establish first the limits of the Paris region, in terms of the grand transportation corridors (Remaury 152). Reinforcing the desire to both know and control the growing region and address the inadequacies of the current transportation infrastructure, his map prominently highlighted new autoroutes and divided highways and clearly delineated – geographically – where the region ends, officially authenticating Paris's newly enlarged shape. Prost's map acknowledged the city's changing suburban identity and recognized the growing presence of the *banlieue* (suburb). His maps followed Jaussely's lead and documented future development and existing green space. But in one significant difference, Prost suggested controlling urban growth. This chapter investigates how Henri Prost's *Carte générale* demonstrates the French government's desire to move beyond the ideals of urbanism, as found in Jaussely's 1919 *Plan d'extension*. Although Jaussely provided new and exciting ideas for the Paris region, Prost provides a much more realistic plan to address the growing region's needs. His presentation of the suburban belt, feared, hated, and named the *ceinture rouge* (red belt) in the 1920s, helped introduce it to many Parisian insiders.

Conclusion: The tumultuous nineteenth century brought Parisian-led regime change in 1830, 1848, and, in many respects, 1870. Although Napoleon III and Haussmann had hoped their works would tame the capital city as they constructed uniform boulevards and transformed the crowded medieval centre into a bourgeois space, throughout the twentieth century, the movement of people and goods throughout the Paris region remained a challenge and official maps showed how to address that issue. The German occupation during the Second World War effectively ended any hope of Prost's 1934 plan coming to fruition. However, the damages inflicted on the city during combat allowed leaders to refocus their attention on it. The pre-war work done by the Service géographique, Jaussely, and Prost allowed future urban officials, such as Lopez and Bernard Lafay, to address problems such as increased traffic, parking, housing shortages, decentralization, and increased sprawl. The end of the war shifted national priorities away from the capital but by the 1950s, economic growth meant that urban planners needed to focus yet again on ameliorating development in greater Paris.

Chapter 1

Working with Maps

> The hidden agenda of mapping, including cartography's modern claim to accuracy of representation, is precisely what makes them interesting and problematic texts, first in the silence of the maps (those elements of the landscape that are omitted) and second (and often related to the first), in terms of the implicit and explicit authoritarian nature of the map as a tool of power.
>
> —Harley, *The New Nature of Maps*, 14

There are both challenges and insights to be gained from working with maps. To begin with, maps are a form of communication, yet each individual reads them in accordance with his or her existing, socially and culturally defined, geographical conceptions (Edney, 'Harley' 48). Therefore, deciphering a map-maker's message and determining his or her audience represents a fundamental task before proceeding with any cartographic investigation. For a nation such as France, with inadequate access to education prior to the Third Republic, establishing who would see and understand a published map marks an important first step in any such investigation. Until Jules Ferry's educational reforms and Vidal de la Blache's efforts added geography to the national curriculum, most citizens had little or no access to – or familiarity with – cartography. The small public who could read and understand maps restricted the medium's reach and effectiveness. Yet, once the French map-reading public started to grow, so did cartography's influence.

Limited map contact mirrored a general lack of national or regional mobility. Throughout most of the nineteenth century, only a very small percentage of the population had the means to board a train and visit Paris. Even short trips were difficult and expensive undertakings (Weber, *Peasants* 198). Factors such as these restricted the formation of a common French identity and slowed any potential effectiveness of national maps, along with any need or expectation to 'see' what France, or any particular region, looked like on a map. Once the Third Republic improved exposure to cartography, however, it could begin to appropriate the medium into its mission of creating citizens and use the science to learn more about itself. This newly acquired knowledge will also provide the necessary tools to view and understand the national terrain.

In addition to understanding any intended or prospective audience, the act of cartographic interpretation itself can be complex and creates its own set of problems. Maps are constructed to stand in for the world in a wide variety of social exchanges: imperialistic, governmental, cosmographical, mercantile, intellectual, and so on. They represent 'reality' – spatial relationships – according to the nature of those exchanges. Even if an individual makes a map as part of his or her personal acts of visualization, the map's structure of space – its very conception of space – and its other representational schemas will be part of the larger ambit of human representations (Edney, 'Harley' 9). In order to recognize or anticipate communications between map-maker and his or her reading public, today's researcher must turn to a variety of tools. The study of semiotics is one such instrument. It asks the reader to take the map apart; to peel away various layers of structure and meaning, in order to uncover possible significations from signs, colour, text, or symbols. Unfortunately, not every map user reads a map the same way its maker intended, unless both user and maker share precisely the same environment (Edney, 'Harley' 48). This consequence represents one of the greatest challenges to map analysis and much can go wrong. Any potential misinterpretation can confuse or distort the cartographer's original message and negate some or all of his or her work.

Conversely, there are also many insights to be gained from working with maps. Cartography reflects a historical society differently than technical, economic, or political avenues. Whereas written documents, transcripts of official legislation, or newspaper reporting deliver precise details of important events, maps provide a more global view and perspective of a past time or space. In addition, cartography provides many layers of textuality: the map itself, the immediate context of the map (its caption, the chapter, and the work of which it is a part) and the wider context of the map (the opus of the individual cartographer or school, the opus to which the text itself belongs, the sociocultural context of the work) (Pickles 50). At first glance, a map indicates what a past or present society considers important enough (or not) to record. In nineteenth-century travel guides, for example, map-makers often silenced undesirable urban areas and their populations by simply leaving them off maps. The refusal to document this space accurately affirmed its lack of value to both the bourgeois editor and reader. Conversely, historical maps can also highlight a regime's agenda. Nomenclature, one such tool in this endeavour, can not only date a map, but also demonstrate which territory, streets, or buildings were important enough to be documented or be given new names following a revolution or regime change.

Cartography as an Urban Planning Tool

Beyond the affirmation of governmental viewpoints and presenting the nation to its citizens, cartography also helped plan Greater Paris. Alphand's 1889 'Les Opérations de voirie,' discussed in Chapter 3, proudly shows what the Third Republic had accomplished in its first two decades, yet it was far from a comprehensive planning document. An officer of the government, Alphand affirmed his ability to continue Haussmann's work and presented the government as both productive and efficient. As the nineteenth century drew to a close, the institutions of police, public health, architecture, and urban planning were all rendered possible by the technologies of political statistics and mapping. In England, Charles Booth's maps of London, for example, helped residents 'see' the effect of crowding in the city's slums. This 'objective gaze' was highly efficient, since it led to a naturalization of the moralizing discourse on poverty (Pickles 131). Paris experienced many of these same social changes too. However, unlike its British counterpart, the French capital had physically marginalized much its poor. When Haussmann razed central Paris's slums to embellish its core, he pushed their residents out to the city's edge and beyond. Squalor, once ever-present in the daily lives of all Parisians,[1] was now mostly tucked out of sight from the bourgeois gaze. By the twentieth century, French officials were ready to acknowledge and improve marginalized

1 Haussmann's *Mémoires* (Vol. III, p. 28) describe his daily walk to school through sordid slums on the Île-de-la-Cité and the 'population qui habitait là!'

neighbourhoods and these once-exiled spaces now appeared on state maps. This cartographic representation, in turn, reintroduced the poorer sections of the city to the bourgeois map reader and allowed him or her to 'see' that population and where they lived and to 'visit' them from a safe distance.

Of course, French urban planning as we know it today did not truly take hold until after the First World War, following in part the influence of the members of the Musée social movement and the creation of the Société française des urbanistes.[2] State mapping prior to these efforts established the foundation upon which the science of urbanism, and a competent map-reading audience, could start and flourish. The maps investigated in this book all present a bird's-eye view of Paris as it developed and expanded throughout the Third Republic. However, unregulated growth in suburban Paris accelerated as peasants left farms for factory jobs in the capital. But urbanists felt that the government should better manage this zone and that state cartography could do more than just document how a space had changed. In the twentieth century, the emergence of master plans was hoped to ensure effective administration and a proper functioning of the city. Administrators struggled to deal with the changing roles of cities and their populations, and visualizing the city and mapping its elements became an obsession of urban planning (Pickles 130). Cartography both showed where the state should focus its energies and presented the changing region to an enlarged map-reading public.

Historically, maps were crucial tools and administrative resources through which the apparatus of the state, whether a republic or an absolute monarchy, exercised its power (Edney, 'Harley' 91). Once France decided to plan the future of the Paris region, officials turned to experts for guidance. These specialists relied in turn on already established cartographic surveys of the region from which to plan Greater Paris's future. Maps play a significant role in how planners modify urban space, even if politicians themselves make the ultimate decisions on what gets built and what remains an idea on paper. People in power produce their dominant reality in any given time and place, so any change to Paris's urban footprint depended on political will.

2 Founded in part from the Musée social's Section d'hygiène rurale et urbaine. A discussion of the growing urbanism movement in France can be found in Chapter 5.

Both government agents and social elites impose their values, but also exercise censorship, over maps. In Paris, not only had the city's poor been marginalized cartographically, but so had women. Barbara Hooper has explored socially produced space that excludes the feminist viewpoint: urban change in Third Republic France – or Paris, more specifically – was 'not plural, but singular – that is male, white, bourgeois.' The practice of governing space often places feminism at the 'periphery and in the process cut itself off from one of its most radical and nurturant roots' (Hooper 48–49). Third Republic cartography should be viewed as a masculine bourgeois product, one produced by the bourgeois political class for a primarily bourgeois audience. Present-day readers should be mindful of this partiality when analysing cartographic discourse. For example, governmental institutions, such as police stations, barracks, and ministries, figure prominently on city maps, while more traditional feminine spaces at that time, such as individual homes, lack the same cartographic recognition.

However, the role of women in the dissemination and assessment of French geography influenced cartography and cannot be disregarded. Augustine Fouillée's *Le Tour de la France par deux enfants* played an outsized part in providing cartographic access to millions of French school children throughout the Third Republic.[3] Later, post-First World War, *géographe* and *bibliothécaire* Myriem Foncin provided valuable, informed critiques of many of the urban plans submitted to the government for the 1919 loi Cornudet.[4] Even though few women published their own maps, their work ensured that French geography remained effective and that it succeeded. Social awareness helps ensure that, when working with historic maps, a researcher better understands how a society viewed and identified itself. After a century of successive revolutions and regime changes, France was ready to promote calm, stability, and growth. Furthermore, it needed to project Paris as its modern, accessible capital city.

3 See Chapter 2.

4 A discussion of Foncin and her critiques can be found in Chapter 5.

The Question of Modernity

David Harvey labels Haussmann as 'one of the great legends of modernist urban planning,' for the *préfet* 'devised a coherent plan to reorganize the spatial frame of social and economic life in the capital' (*Modernity* 111). Haussmann's modernity thus changed how politicians and bureaucrats viewed and manipulated urban space moving forward. Of course, there is no one straightforward definition for 'modern,' nor is there a single event that can be used to determine the arrival of 'modernity' in France. Under the heading 'Two Modernities,' Matei Calinescu discusses an 'irreversible split' between modernity as a stage in the history of Western civilization and modernity as an aesthetic concept (41). Post-1871 leaders needed to convince the bourgeoisie that Paris had finished with rebellion and that technology, combined with enhanced transportation, would improve daily lives. Through maps, politicians could influence how the city was portrayed and what messages were transmitted. While aesthetic modernity was brought into being by the avant-gardes and expressed itself through diverse means, ranging from 'rebellion, anarchy, and apocalypticiscm to aristocratic self-exile' (42), Calinescu's term – modernity as a bourgeois ideal – emphasizes the power structures involved in finance, industry, government, and publishing. In this definition, he highlights

> a doctrine of progress, the confidence in the beneficial possibilities of science and technology, the concern with time (a measurable time, a time that can be bought and sold and therefore has, like any other commodity, a calculable equivalent in money), the cult of reason, and the ideal of freedom defined within the framework of an abstract humanism, but also the orientation toward pragmatism and the cult of action and success – all have been associated in various degrees with the battle for the modern and were kept alive and promoted as key values in the triumphant civilization established by the middle class. (41)

In this respect, modernity permitted a larger portion of the population to take advantage of the scientific and technological advances made throughout the nineteenth century, thus directing economic largesse into the pockets of the middle class. Economic expansion, as manifested in spectacles such as Napoleon III's Expositions Universelles of 1855 and 1867, both of which were housed in the aptly named Palais de l'Industrie, adopted the bourgeois modernist ideal to showcase Paris as a cultural and international destination. Both foreign and provincial visitors to these staged events were eager to come and see what the city had to offer. By this definition, modern Paris was a success.

Anthony Giddens asks, 'what makes one society more modern than the societies of [...] forbears?' For him, modernity is 'an historical condition of difference, in one way or another a displacement of everything that has gone before' (15). Supplementing both Calinescu and Gibbens, David Harvey focuses on modernity from a French perspective. He suggests that '"one of the myths of modernity is that it constitutes a radical break with the past" in that "[it] is ... always about 'creative destruction'"' (Harvey, *Modernity* 1). Initially it appears that pre-Third Republic, Baron Haussmann established this 'radical break' and effectively separated the Second Empire from the unpopular July Monarchy and short-lived Second Republic.[5] During Haussmann's tenure as *préfet* of the Seine, Parisians saw their capital radically transformed from a city many from the Middle Ages would recognize to one that was a model of bourgeois ideals, covered with wide boulevards, provided with large supplies of fresh water, and furnished with factories, slaughterhouses, and a new sewer system that, Haussmann proudly thought, had no equal in Europe. For Haussmann, being a modern Parisian meant creating a secure, orderly, and clean city. The *préfet* believed that a more strategic embellishment of widened streets would render any revival of the 1848 barricades impossible, and with the emperor he planned many new straight avenues to connect military barracks with workers' districts. A modern city would also be trouble-free and the Paris of the late 1860s, pre-Commune, believed itself successfully disjoined from its violent past. Thus, in the 20 years of Napoleon III's empire, city space changed from one crowded, dirty, and weakened by rebellion to one designed for pleasure and business, and maps of the Second Empire eagerly promoted this new reality.[6]

5 It should be noted, however, that Harvey also suggests that much of what Haussmann did was already present in embryo form in the 1830s and 1840s. At the same time, Harvey argues that Haussmann contributed immensely to how the Saint-Simonian notion of modernity was articulated (85).

6 Guidebooks, often the most accessible maps for the public at large, highlighted bourgeois playgrounds in the city. The Bois de Boulogne, the Parc des Buttes Chaumont, wide boulevards, or the Exposition Universelle

However, looking for their own 'radical break' and unwilling to endorse Napoleon III's Second Empire, Third Republic administrations viewed modernity as a complement to republican values. As with their predecessors, they too documented the Republic on maps, using cartographic discourse to emphasize the regime's triumphs and to downplay its shortcomings. The cartographer J. B. Harley defines map discourse as 'those aspects of a text which are appraisive, evaluative, persuasive, or rhetorical, as opposed to those which simply name, locate and recount' (54). Used to unite as well as to support, maps validated the new democratically elected regime of the Third Republic. Completed projects, schools, trouble-free neighbourhoods, and numerous Expositions Universelles commonly appeared on maps sanctioned by the government. The map reader would see only the best of Paris. Anything that might contradict this message was eliminated, simply removed the sight.

Framework for Map Image Analysis in this Book

Throughout the nineteenth century, French cartographic history had been influenced by the social and political forces that helped to encourage map use. Where 'l'histoire du XIXe siècle s'intéresse très peu aux images, et encore moins à cette catégorie d'images que sont les cartes' (Weber, 'L'Hexagone' 1178), there was increased exposure to images in the form of cartography in Third Republic schools. Generally, early nineteenth-century geography textbooks, such as Joseph Gibrat's *Traité de la géographie modern* (1813), did not contain any maps. But by the beginning of the Third Republic, maps start to emerge in the educational system. The military defeat of 1871 initiated an internal examination of what French school children learned, and it was noted that their lessons often favoured history over geography.[7]

site often overshadow residential neighbourhoods and industrial quarters. See Adolphe Joanne's 1867 *Paris Illustré* guidebook for the city.

7 The notion that France had been defeated in the Franco-Prussian War by an army that was better prepared, better equipped and, above all, better educated than the French forces was widespread at all levels of public discussion, from the popular press to the flood of pamphlets and the debates at the National Assembly. Many called for maps to be made more available and brought up to date, not just for the army, but in schools too (Andrews 177–78).

With little understanding of how to read maps on the battlefield, one consequence of this defeat was a re-examination of the role and subsequent inclusion of geography in the national curriculum. The result: a new relationship between the French and maps; a transformation which served many purposes. By the First World War, maps were far more prevalent and appeared in all aspects of daily life. Smaller, specialized ones, in the form of travel and transportation guides, for example, provided users of urban public transportation of all ages with plans of rail lines and omnibus routes and, for the first time, urban spaces. Upon reading them, it became much easier to visit, and therefore know, many more neighbourhoods of a city. This exposure to geography and expanded transportation options widened the French map-reading audience considerably as these users of public transportation saw and used cartographic representation daily.

This increased cartographic availability throughout the Belle Époque helped eliminate the geographical unknown but also strengthened French notions of specific physical locations, such as the city limits of Paris or what France as a whole looked like. Besides opening up new avenues of knowledge, what is the primary function of maps? They represent. They represent not only familiar cities and areas, but also those that we do not know and about which we wish to learn more. The map is the apparatus that shows what no eye can see, even when it represents recognizable territory. Maps give as much to thought as to sight; they project order and reason on the world (Jacob, *L'Empire* 15–16). Even today, map readers are often willing to accept maps as unbiased informational sources. As readers, 'we're accustomed to thinking of maps as useful objects, whose purpose is to direct us from one location to another' (Lehman 131). Nevertheless, most cartographers will agree that there is much more to them. Maps are never value-free images, and they are never independent of the arrangement of power in society (Harley 53). The government wished to transmit its view of the changing city. Maps' presentation would comprise the ongoing Third Republic narrative of how to define Paris.

How, then, to examine maps in this book? The flow of cartographic information begins with its selection by the cartographer from geographical reality. Next, he or she guides the map's 'metalanguage' towards the map user's conception of geographical reality (Edney, 'Harley' 39). To be successful, maps need to portray the world in a manner that provides detail, but does not confuse. In France, cartographers presented city space in a way that would highlight the new republic. Through the investigation of map discourse from Third Republic France, including a deeper understanding of the desired communication between the cartographer and the map reader, we can evaluate and identify what the government wished to promote, and establish interpretive frameworks for the maps studied in this book. The reliance on symbols, shading, text, and even silence helped maps define Paris as a modern, secure, growing, democratic capital city eager to accept both business and tourism. Yet this historical map discourse also provides clues to understanding how the reader imagined urban space in terms of both modernity and mobility. How do maps reflect the perceived or imagined definitions of Paris? How did cartography help transmit understanding of the Third Republic's capital city to readers?

Maps have embodied power throughout history, yet applying the science to historical images involves examining subtle messages and map discourse. All maps possess a deeper, perhaps hidden, level of symbolism that communicates socially, culturally, and politically charged understanding of spatial significance (Edney, 'Harley' 85). Since the present-day researcher works without absolute knowledge of a past author's intentions, he or she must decode the meaning of his work. Semiologist Martine Joly explains further:

> Interpréter un message, l'analyser, ne consiste certainement pas à essayer de retrouver au plus près un message préexistant, mais à comprendre ce que ce message-là, dans ces circonstances-là provoque les significations ici et maintenant. Il faut en effet et bien entendu des garde-fous et des points de repère à une analyse. Ces points de repère, on pourra précisément aller les chercher dans les points communs que mon analyse peut avoir avec celle d'autres lecteurs comparables à moi. Certainement pas dans d'hypothétiques intentions de l'auteur. (36)

Therefore, through a methodical semiological analysis, one can hope to better understand the author's point of view.

The first strategy in map analysis for this book is the attempt to identify 'the rules of social order' within the map (Harley 45). Every map manifests two sets. First, there are the cartographers' rules, which operate in the technical practices of map-making. The second set involves external factors, which 'can be traced from society into the map, where historical factors influence the categories of knowledge' (Harley 45). In Third Republic France, these standards included access to cartography and geographical knowledge. As such, the map becomes a 'signifying system' through which 'a social order is communicated, reproduced, experienced and explored' (Harley 45). Historical maps combine these rules to create an interpretation of reality – the map-maker's reality.

An accurate comprehension of historical factors is paramount, but equally important is knowing how to apply map-reading techniques to specific texts as they contain a system of signs. The map-reader's eye first targets the most noticeable objects on maps: 'Je vois du rouge, et pas du vert, ni du bleu, ni du jaune, etc. [...] Je vois un cercle, et pas un triangle, ni un carré, ni un rectangle, etc.' (Joly 43). The first preliminary scan determines the tone of a map. Symbols and shapes chosen by the cartographer 'replace' text, which creates a visual message comprised of different signs. 'On peut admettre qu'une image constitue en effet toujours un message pour autrui, même lorsque cet autrui est soi-même' (Joly 45). Everything present on the map has been explicitly placed there by the author. For example, once Paris wished to improve access to the city, roads figured prominently.

Finally, when distinguishing between a visual message's recipient and its function, there must be certain standards. According to Joly, referencing Roman Jakobson's *Essais de linguistique générale*, when effectuating an analysis,

> deux méthodes se proposent à nous et peuvent se montrer opératoires: la première consiste à situer les différents types d'images dans le schéma de la communication. La seconde à composer les usages du message visuel à ceux des principales productions humaines destinées à établir un rapport entre l'homme et le monde. (46)

These reference points situate different types of information and confirm the relationship between man and world. How can one 'represent' the Republic on a map? Map discourse is an important tool for better understanding an historical society. Map discourse never involves simply naming or locating a feature on a map, but calls on a reader to consider the political significance of every feature included. Map rhetoric is subject to a cartographer's ability to represent and map symbols; drawings, or images more generally, are never more than representations of the map elements themselves.

Since maps are comprised mainly of symbols and abbreviations. 'Signs,' 'symbols,' and 'icons,' terms employed by Charles Peirce, help contribute to a full semiological understanding of cartographic documents. Peirce defined a 'sign' as 'a thing which serves to convey knowledge of some other thing, which it is said to stand for or represent' (13). So something as simple as a name or label representing an object can serve as a sign on a map. To this Peirce added 'icons,' which are 'signs that stand for objects,' and most, 'if not all, are likenesses of their objects' (13). Examples might be employing the outline of a factory to represent industrial plants, as in the 1906 *Carte de France*, or using a red cross for a hospital, as Jaussely did in his 1919 *Plan d'extension*. Peirce's third term, 'symbol,' is 'a sign … represented by an icon associated with it' (17). Some of the most common 'symbols' in this regard today might be the use of a cross for a church, or a bell or apple to represent a school. Map colour, shapes, symbols, legends, titles, and scale, individually or together, incorporate Peirce's semiotic terms and are necessary in determining the discourse 'chosen' by the cartographer and publisher, helping to 'establish a connection between mind and object' (Peirce 14).

Barthes's *Mythologies* took Peirce's semiotics one step further by examining contemporary objects as myths. By breaking down the discourses behind the myths of maps, more specifically those printed in the latter half of the nineteenth century, one may better determine what, if any, message was being promoted. It is possible to apply to maps Barthes' rhetorical figures like 'tautologie' (226), in which the status quo is the primary emphasis. Perhaps maps belong to a 'privation d'histoire' (225), legitimizing the government in place and trying to suppress any previous or present history or story of how the city of Paris was transformed. Or perhaps maps fit better with another of Barthes's rhetorical classifications, 'identification' (225). 'Impuissant à imaginer l'Autre,' such as the dangerous or working-class sections of Paris, 'le petit-bourgeois s'aveugle, l'ignore et le nie' (226). Just as the petit-bourgeois chose not to acknowledge that which he didn't know, maps were transformed into or masked by something the reader did not fear, for example either a shaded-in city block, an oversized monument, or the use of a bright colour. Alphand's red and yellow roads in Paris's outer *arrondissements* may distract readers from noticing any nearby shantytowns. Barthes concludes that images are constituted by an architecture of signs drawn from a variable depth of meanings. Each image, no matter how complex and layered the meaning, is still being coded.

Republican iconography was coupled with documented modernity and found its way onto maps. Educational reforms created a demand for new schools; urban security involved increased police presence, in the form of *casernes* and prisons; a growing industrial base necessitated the construction of factories, many of which appeared on Third Republic maps. Paris had changed during the nineteenth century. It became a republican, industrial city, and government maps reflected this transformation.

Apart from maps, print more generally has always provided an excellent window onto past and present societies. Newspapers, magazines, and popular literature have long been analysed to determine how a public perceives both society and current events.[8] To this viewpoint, maps provide their own crucial, if biased, impressions. 'While the object of mapping is to produce a "correct" relational model of [a] terrain, its assumptions are that the objects in the world to be mapped are real and objective' (Harley 154). But, in many cases, they are not. Maps are pre-eminently a language of power (Harley 79). Their existence can be used to promulgate regimes in power, favour an ideology or sell a product. They are a socially conservative vocabulary used to favour the status quo (71). Successful revolutions change maps; failed ones lose map space and standing and then disappear into cartographic nether regions.

For Third Republic France, the status quo meant promoting republican ideology. However, successful cartography strives to be helpful, attractive, and welcoming. Regardless of ideology, maps must invite a reader's attention. The cartographer Christian Jacob proposes two *lignes de forces* explaining the attraction to maps. First, he identifies a powerful, imaginary draw by which the map invites us to look, explore, and dream. Second, he labels maps as an example of rational construction. We fill space with our knowledge of geometry and geography (16). The map is a model used to interpret and interrogate as much as simply to read. Michel de Certeau explains the lure of

8 For a more in-depth examination of the role of the press and pre-twentieth-century print culture, see Darnton's *Revolution in Print* and Cate's *Prints Abound*.

maps by mentioning the view from above, which provides a heavenly gaze. Surveying New York City from the hundred and tenth storey of the World Trade Center, Michel de Certeau explained a disconnect between cartography and reality as he looked down on the island of Manhattan. From that vantage point, he saw how 'l'agitation en est arrêtée […] la masse gigantesque s'immobilise sous les yeux' (139). From high above the street, the city's notoriously frenzied activity seemed well under control, the way planners like to 'see' any city. This version of New York is a simple collection of skyscrapers and long avenues. De Certeau contrasts that view 'd'en haut,' where 'le corps n'est plus enlacé par les rues qui le tournent et le retournent selon une loi anonyme' with the view 'd'en bas,' 'où vivent les pratiquants ordinaires de la ville' (140–41). On the ground, Manhattan has a very different feel. The noises, smells, and crowds of people overload the senses, distract the urban wanderer, and render the city unintelligible. De Certeau elaborates on this divide, proposing that 'le marcheur constitue, par rapport à sa position, un proche et un lointain, un ici et un là' (149). The 'ici,' what pedestrians experience daily, never reconciles perfectly with the 'là,' the more abstract, mental notion of space. It appears that the most helpful urban solutions must merge these two ideas into one effective plan. Various planning authorities needed to know the terrain of the Paris region, not just modify it. For De Certeau, the map reader assumes a powerful position, viewing and imagining the city in a way not possible on foot. This view from above has traditionally been a domain reserved for urban planners and cartographers. Today, however, anyone with access to the Internet can now experience the once-elusive heavenly gaze. At the same time, the map encourages interaction between the reader and the document. You can touch it, use your finger to follow streets, mark a path, or simply fold it up and place it in a pocket for easy access if you suddenly lose your way. This medium allows for a more active role than traditional narrative.

Historical French Cartography

Maps are also political instruments. Through them, leaders have clearly documented their power and control for centuries. City plans for Rome and Constantinople, two such examples, were found in the library of Charlemagne (Thrower 41). During the Middle Ages and the Renaissance, maps figured prominently in both the imagination and propaganda of European rulers. A highly stylized map served both to adorn, for example on a walled tapestry, and to document newly acquired territorial gains, often with stylized decoration substituted for accuracy. In the fourteenth century, European exposure to the Byzantines and other cultures introduced new cartographic techniques that enhanced accuracy, which broadened its appeal. Upon Europeans' arrival in the New World at the end of the fifteenth century, maps served an additional purpose: as informative devices. The documentation of newly found territories provided legitimacy, affirming both the extension and influence of the European crowns and the Catholic Church. European 'acquisition' of territory through cartographic nomenclature became a common practice (Thrower 67). Large, mapped territories of the New World provided Spain, Portugal, and soon other nations with colonial empires, new subjects, and bragging rights on the diplomatic stage.

The absolutist leaders in early modern France continued this phenomenon. In the field, military victories signalled that France was becoming larger, stronger, and more important in Europe. On paper, Richelieu eagerly used cartography to record these gains. The cardinal continued the French state's support of the work of geographers and cartographer-engineers, enthusiastically embracing the use of maps in the formulation of state policy and defence (Vayassière 252). Acutely aware of the strategic importance of geography, Richelieu undertook numerous projects involving the exploitation of physical space, including the continued fortification of the nations' borders, the annexation of adjoining territory, and the expansion of the king's colonialist designs (Peters 26–27). Under Louis XIV, in a period in which no global knowledge of the French state was available, Colbert's demands for corrected maps were forwarded to Nicolas Sanson, the *géographe ordinaire du roi*, for examination. This process helped cartography, a language through which the king's efforts to observe the entire country from his central position of symbolic power in Paris, and later Versailles, assume material form (Peters 78).[9] In seventeenth- and eighteenth-century Europe, for example, maps were often published

9 For a better understanding of French cartography, see Tom Conley's *The Self-Made Map*, Pierre Pinon's *Les Plans de Paris*, and Antoine Picon's *Un atlas parisien*.

for rulers to document authority or serve as decoration. The use of maps changed as quickly as did the technology employed to print them. In France, the rise of absolute power under Louis XIII and Louis XIV funnelled resources into French map production. This allowed France to overtake the Low Countries as the centre of cartographic production in Europe (Tooley 40). Significant royal support for cartography continued into the eighteenth century, when French influence on European geography soon reached its height.[10] Triangulation, a system of mapping ground coordinates based on astronomical observations, started under Louis XIV. This technology allowed map-makers to correct previous cartographic works, at which point L'abbé Picard, head of the royal cartographic operation, produced a new accurate map of France. Picard organized a team for measuring France using self-designed instruments: 'Picard mesure avec un soin extrême la base d'une chaîne de treize premiers grands triangles: c'est le premier tronçon de méridienne prolongée sur les platitudes d'Amiens en 1670, puis étendue après 1700, et terminée en 1718.' With this method, Picard drew the first 'corrected' (i.e. smaller) map of France, causing Louis XIV to proclaim, 'Ces messieurs de l'Académie, avec leurs chers travaux, m'ont enlevé une partie de mon royaume' (Vayassière 255). By the eighteenth century, French geography was at its height. The Cassinis, one of the best-known families of French royal cartographers, started triangulating a monumental *Carte de France* in 1744 and turned in the first of 180 sheets to Louis XV in 1756.

The Cassini family is an excellent illustration of Ancien Regime map producers in France. Jean-Dominique Cassini (1625–1712), like most cartographers of the period, was a scientist. Earlier in his career, he had surveyed waterworks and fortifications in Italy. His 1669 publication of tables of the eclipses of Jupiter's four major satellites brought him to the French king's attention in Paris. Cassini had not wanted to stay in Paris for more than a few years, but Colbert encouraged him to remain to work on one of the king's pet projects,

10 French supremacy in so many of the arts was so great that France took the lead from Holland not only in cartography, but in geographic science too. It was in France that scientific mapping from exact ground observation began, and speculative cartography finally abandoned. Specific examples of this transformation include Nicholas Sanson's *Atlas universel* and R. J. Julien's *Atlas géographique et militaire de la France* (Tooly 40–43).

demonstrating France's superiority over other smaller kingdoms by continuing the mapping of France (Wilford 132). Cassini stayed and started one of the best-known French cartographic dynasties.[11] Later, separate from astrology, cartography became a career in its own right. At the height of the French Enlightenment, the publication of Cassini's *Carte de France* (1744) epitomized continued royal support of cartography as an essential science in the eighteenth century. The state maps of eighteenth-century France were intended for public consumption as well as official use. At different times, however, their development depended on private patronage beyond the sphere of the state in a nascent public, and 'social,' sphere. These examples can be contrasted with the Hapsburg (quasi-dynastic and 'police') maps of 1774, only three jealously guarded and hand-drawn copies of which were kept by the state (Joyce 39).

In Third Republic France, government continued to use maps to promote both France at home and colonization abroad, the latter of which seemed especially important after the loss of Alsace-Lorraine. Eugen Weber suggests that after 1870, 'la carte du nouvel Empire français, ou d'un monde sur lequel l'Empire s'étale, vaste et rassurant, accompagne désormais, et parfois éclipse, celle de la France diminuée' ('L'Hexagone' 1184). Eager to promote republican values and French aims, the Third Republic looked to foster a French national identity through government programmes, national elections, and now maps, even if they showed negative information. The post-imperial government hoped that a damaged France documented though maps would be easy to understand and that visualizing the missing territory would lead to a better grasp of the French defeat. Citizens would therefore be more receptive to avenging the loss of 'les provinces perdues' to the Germans.

No longer predominately government instruments, today maps are now accepted as regular everyday objects. Yet throughout history, cartographers have always had a special relationship with power. Education, the cost of materials, and the time needed to colour maps historically limited their use to the Church and nobility, and this relationship to power added to their legitimacy (Ehrensvärd 127). Through its association with authority, a map becomes a

11 His grandson, César-François Cassini, emerged as the most persistent, accomplished map-maker in this illustrious family. In recognition of his many achievements, he was made a count, Cassini de Thury (Wilford 138).

multilayered, accepted representation of a physical area. Consequently, by documenting a society or a city, the published map becomes fact, which the reader willingly accepted. This recognition of a map's authority cannot be ignored.

Cartography as a Reflection of a Society

Due to educational reforms and travel, by the start of the twentieth century more French citizens were seeing more maps of more destinations. Exposure to cartography is easily explained and documented. For the nation, the nineteenth century had been an integral period in the histories of both cartography and print culture more generally. People's relationship to maps changed considerably as increased exposure to them meant greater familiarity and ease in use. In addition to semiotics, a better understanding of the history of cartography, especially of printing and colour, may also help clarify map reception. To this end, historical cartography finds itself dealing with the challenging task of explaining the role of maps in past societies. While some past maps are explicit in their purpose, such as those documenting a voyage by an explorer or the results of a peace treaty, others are more opaque, with no overt message, and thus require a more in-depth study. Printed texts in Ancien Régime France, including maps, were not used or read by a mass audience due to low levels of literacy and high production costs.

The major map-making projects of the seventeenth and eighteenth centuries undertaken by the Cassinis and other cartographers dealt primarily with the kingdom as a whole. In general, city maps were not commissioned except when making large-scale urban planning changes and then only primarily in larger cities like Paris, Lyon, or Bordeaux. Other forms of urban mapping did exist, but on a smaller scale and often as part of larger territorial maps. Denise Turrel's *La Couleur de la ville* (1998) examines the representation of cities in regional and national maps of the Ancien Régime. Icons such as castles or churches used to symbolize urban settlements demonstrated royal authority. In the seventeenth century, maps rarely included legends, so the cartographer relied almost exclusively on symbols to do the work. If any type of legend did exist, it included only simple *explications* of urban symbols, documenting a limited progression of value. Adding a building for each category of construction culminated in distinguishing the city, 'un lieu plein de maisons' (Turrel 128), as the summit of all French communities. This representation of larger inhabited areas not only signified a larger potential literate population but, perhaps more importantly, also indicated where most people who could use maps resided. For these reasons, cities often stood out from the surrounding countryside.

The evolution of French cartography and its associated printing technology also affected the final product. Colour, shape, and size all determined how a reader experiences a map. Yet when it comes to promoting cartographic discourse, how might a late nineteenth-century cartographer highlight their ideas and subtly (or overtly) favour a republican ideology as discussed in later chapters? Size has always been an effective way to document hierarchy, as we witness in Fouillée's map text in *Le Tour de la France par deux enfants*, yet colour remains another useful technique to demonstrate partiality. For maps without detailed legends or descriptions, colour is one of only a few indicators of importance and it often defined a city, helping it stand out on a larger map. Furthermore, if cartographic symbols and shapes appear ambiguous to the reader, either through poor labelling or lack of text, colour becomes even more essential to providing meaning and assigning value. Throughout history engravers often fortified their representations with biased, loaded hues, which at the same time proclaimed that the place in question was of value (Turrel 128). Prior to the nineteenth century, map-makers often employed a reddish crimson to represent cities on maps, a convention which provided a *de facto* standardized urban representation. This homogeny helped any able map reader to better understand urban map markings, if not specifically labelled as such, as that colour referred to cities.

Even though one can look back and recognize significant uniformity, cartographers caution that after the fact it is too easy to assign 'importance' to certain colours. For example, certain tints, such as the crimson found in Ancien Régime maps, should not automatically signify representations of power, even if there are examples that suggest the contrary for at least two reasons. First, while geographers and cartographers constructed maps, their work was often later interpreted and published by a separate printer or engraver. On one map, the map-maker may have specifically chosen red to represent power or a city. On a second map, however, we may never know if that same tint was actually the cartographer's choice or just the result of circumstance, such as a printer's ability (or lack thereof) to produce any other colour at the time of publishing. Therefore, any dominance perceived today on pre-nineteenth century maps is more likely not the result of a choice by the person who drew the map, but he who printed it. Second, even if the geographer was actively involved in the printing and publication of his map, the colours choices were often aesthetic:

> Les géographes en effet ne pensent pas leur carte en couleurs, mais essentiellement en noir en blanc. Pour distinguer les catégories de villes, ils échafaudent des systèmes qui reposent sur des symboles ou des règles d'écriture des toponymes, à destination des graveurs. Mais leurs traités ne proposent jamais un code de couleurs pour les enlumineurs. Pour eux, les coloris ne sont pas liés à la variété de la carte, mais à sa beauté – ainsi Pierre Duval oppose-t-il nettement le savoir-faire artistique des enlumineurs à leur manque de connaissance. (Turrel 131)

In an era when maps adorned castle walls, this makes sense. By the nineteenth century, however, the printing process had advanced considerably, which meant that the choice of red truly belonged to the cartographer.

Throughout much of history, maps were considered both works of art and cartographic documents. Nevertheless, as technology facilitated the addition of multiple colours to the printed page, it also changed how readers viewed them. For many, a coloured map was 'en effet un objet à voir' compared to a black-and-white map, 'qui est à lire' (Turrel 132). Unsurprisingly, this preference for hand-coloured over black-and-white maps continued for centuries. Regardless of previous obstacles, over time a preliminary standardization emerged which allowed cartographers to present a more uniform sense of what maps should look like and which colours to use. Harley argues that the printing of maps led to the 'technologizing of the map' so that 'map images acquire a tidiness and inevitability lacking in the manuscript age' (98). Yet, even earlier, the eighteenth-century French geographer Robert de Vaugondy cited the need for order in mapping, stating that 'la carte de France était un vrai chaos, qui ne présentait qu'un amas de villes entrecoupé de noms de provinces' (Turrel 133). Where crimson commonly documented seventeenth-century urban development, red began to emerge as a default appropriate representation of brick cities.[12] However, this association was short-lived and towards the nineteenth century the convention progressively diminished. Louis XIV inherited a city (Paris) of brick and left it marble (Horne 125), and other French cities underwent similar urban transformations. A red tint no longer accurately represented building materials and on many maps, cities began to revert back to a more realistic grey or black. However, other parts of the map started or continued to receive colour: water remained blue and parks and forests maintained their green appearance.

Even with ample documentation, precise reasoning behind historical colour use is still quite difficult to verify. Many cartographers today express frustration with trying to decipher printed colour as a determinant in map reception. Map meaning can change and often depends on the circumstances and condition of map readers and interpreters (Edney, 'Harley' 106). Can we truly know if the nineteenth-century map reader reacted differently to mapped colour representations of a city from past readers? The cartographer Karen Ehrensvärd asserts that, throughout history 'colour has clearly developed into an independent means of cartographic expression, albeit one whose visual impact on the map user has yet to be fully understood' (19). How then did the progression of seventeenth, eighteenth, and nineteenth-century map printing techniques, such as the inclusion of a greater variety of colour, affect map reading? Are we able to determine a past reaction to a published map? If so, how? A degree of familiarity with maps, for example, helps researchers today understand how a reader used and reacted to them. Improved access to geography ensured that generations of French better understood maps. Furthermore, an in-depth knowledge of the printing practices and advances in the nineteenth-century French publishing industry clarifies the socio-historical factors involved in interpreting map acceptance. This integral component, part of the ever-changing field of the history of print culture, affects not only historical map reception, but also comprehension. By examining publishing improvements along with the social history surrounding the exposure to maps, we can better comprehend

12 The cartographer Remi Caron mentions how 'la carte est le regard qu'un cartographe et une société posent sur le monde' (*La légende*, 329). Turrel states that it is the map-makers who make the world intelligible by, for example, presenting 'frontières vertes, fleuves jaunes ou roses,' none of which corresponds to reality (133).

what influenced nineteenth-century map construction, as well as how Third Republic readers responded to map size, colours, and text.

Colour Techniques

Throughout the history of cartography, multicoloured map printing has been limited by the availability of techniques, tools, and materials. Restricted by technology, the capacity of map-printing procedures has been the controlling factor on the final map product throughout the history of cartography (Ehrensvärd 124). The actual colouring process itself, has changed considerably over time. Throughout the seventeenth century, hand colouring was a common adjunct to the cartographic enterprise. Until printing and engraving became widespread, the cartographer had control over how his map appeared and held considerable influence over the printer as the final form took shape. In that same century, the colouring of maps, which were no longer reserved for academics or professionals, became an accepted genteel pastime, as indicated by the numerous treatises intended for the instruction of amateurs.[13] However, colouring by hand was slow and if the map-maker needed multiple copies, there was little hope of standardization. Furthermore, the high cost of professional hand colouring limited the distribution of maps and discouraged large-scale production (Ehrensvärd 139). Therefore, any subsequent copy could differ significantly from the original, a less-than-ideal outcome in an industry that relied on precision and data. These potential circumstances add yet another level of difficulty to the examination of historical maps, as they eliminate any common discourse over a large number of hand-coloured maps, even if one model was generally followed.[14]

Throughout the eighteenth and nineteenth centuries, the book printing industry continued to make further advancements from which

13 Many treatises on the art of colouring (first published in Italy and later in England, Holland, and France) traditionally included chapters on colouring fortification plans and maps (Ehrensvard 134).

14 One of the passages already quoted from John Smith's treatise has shown how colour was employed to distinguish different map features as well as in decoration. By the end of the seventeenth century, several map-makers reacted against the distracting use of excessive colour on maps and advocated reducing colour to a practical minimum (Ehrensvard 137–38).

cartography benefited. In colour printing, transfer methods could be employed to make precise guide images for the registration of plates (the correlation of colours in a single image) that were used to put multiple colours on a page. Other significant improvements in that process came from the development of various mechanical devices such as the registration frame for lithography in 1836 and Charles Knight's 1838 system for wood engraving (Ehrensvärd 132).[15] During this period, increasing mechanization made accurate plate registration much easier, a step that helped standardize map production, an overall goal in an industry that lacked it. Now, the cartographer's voice could be seen on every copy.

Yet problems emerged when users tried to adapt the technology to larger scale projects:

> When Victor Raulin's geologic map of the Paris area was lithographed in colour by Kaeppelin in 1842 by overprinting four plates to produce eleven tints, it was heavily criticized for the colours' lack of transparency. The solution soon found was to employ a variety of methods for breaking up the solid colour surface into patterns of small dots, lines, or other marks whose cumulative visual effect was that of a "transparent" colour surface. Trial and error was then needed to find a solution. A technological breakthrough came in the 1860s, when shading on grained transfer paper (coarser in texture than grained stone) made it possible to transfer crayon to the long-wearing smooth-surfaced stones used in printing lithographic ink drawings and lithographic engravings. (Ehrensvärd 141)

Over time, lithographers overcame most of these technical difficulties and achieved a greater standardization in an industry that relied on accuracy.

As geography's accessibility and presence grew in Third Republic French society, different genres of maps materialized, all of which

15 In 1838, Charles Knight received British patent no. 7673 for 'Improvements in the Process of and in the Apparatus used in the Production of Coloured Impressions on Paper, Vellum, Parchment, and Pasteboard by Surface Printing.' Knight referred to the process as 'illuminated painting,' and invented it for the economical printing of coloured pictures, maps, and drawings. For details of this process, see Burch and Gamble, *Colour Printing and Colour Printers*, 141–43).

benefitted from using printed colour. As early as the 1840s, the traditional employment of stencil colour became inadequate for the growing production needs for most cartography. This factor spurred experimentation with additional methods of colour printing, which was vital to the scientific community. Nevertheless, despite its drawbacks, the hand colouring of maps did not die out completely. The original method offered many benefits than the purely mechanized ones. While printing techniques improved via the construction of larger and more elaborate printing presses, hand colouring retained a certain simplicity. It did not require complicated machinery and could be done by almost anyone, anywhere, with the proper colour pigments, supplies, and training.

As cartographic printing became more common, black-and-white maps such as those found in *Le Tour de la France par deux enfants* (discussed in Chapter 2), the easiest and least expensive to publish, became more uniform. But colour continued to gain followers too. In the late nineteenth century, individual topographic maps appeared in multicoloured versions, though large-scale coloured atlases such as Adolphe Alphand's *Les Travaux de Paris* (whose 'Paris en 1889, indiquant les opérations de voirie exécutées de 1871 à 1889' is discussed in Chapter 3) were still a novel concept to many French people. But the advances in colour printing technology discussed in this chapter allowed more multiple-page colour map collections, travel guides being an example, to emerge. Although rare before the Second Empire, it was not unheard of for Belle Époque guidebooks to include at least one coloured map (usually as an insert), even if the vast majority of its other engraved images and maps remained black and white. Adolphe Joanne's *Paris Illustré* (1870), one such example, included a large fold-out black-and-white map of Paris fastened inside the back cover. In the middle of the tome, however, publisher Hachette included many coloured inserts for various central neighbourhoods, such as the Champs-Élysées, Les Halles, and the Champ-de-Mars.

As the nineteenth century drew to a close, large-scale hand colouring on maps all but ended. Printed colour offered many advantages over traditional, hand-colouring techniques, which were no longer uniform enough for commercial or industrial use. As map-makers made the shift to industrial and commercial printing, they did not look to alter already established cartographic norms significantly. In fact, many tried to imitate the preferred, centuries-old hand-coloured look. In order to do so, cartographers kept the same highly standardized associative symbolism (such as blue for water, red for towns, green for parks or trees, etc.), and thus the natural world has come to characterize the representation of the most common map features. Even today, cartographers looking for the closest interpretation of reality possible attempt accurate portrayals through associative colour.

In addition to colour, major improvements were also made in lithographic techniques and in other relief processes, such as wax engraving and chemotypy, which succeeded wood engraving and altered the actual physical printing of maps. By the late nineteenth century, the application of photomechanical methods in printing had made it possible to reproduce an ink drawing without having it recopied by an engraver or a lithographic artist (Ehrensvärd 140). Many of these techniques placed the cartographer in direct control. This standardization saved time and reduced chances of error. Mass-produced, accurate maps provided the French map-reading public with a uniform product. So people saw the same maps with the same colours, font, and images.

Maps as Part of French Cultural History

Under Napoleon III, the promotion of Paris and the imperial regime had been facilitated through imperial propaganda, such as the inauguration of the boulevards. The Third Republic too promoted its own ascendancy by manipulating major public events. Just as had been done with the Avenue de l'Impératrice, Napoleon III's showcase Avenue Napoléon was rechristened as the Avenue de l'Opéra in time for the 1878 Exposition Universelle.[16] The Third Republic also exploited the rise in publishing and travel, and could promote the Republic through travel guides and official maps of the period.

The end of the Second Empire and the beginning of the Third Republic coincided with a century-long growth in the circulation of printed materials. Average daily readership of Parisian newspapers rose from around 50,000 in 1830 to over 700,000 in 1869. This growth

16 In April 1876, the government announced that an Exposition Universelle would be held in Paris in 1878. The *préfet* and the city council agreed that the Avenue de l'Opéra would have immense prestige value if it could be completed on time (Sutcliffe, *Autumn* 51).

reflected not only better literacy rates but also falling production costs, and thus retail prices (Price 260). The rise in rail travel, accompanied by the proliferation of travel guides, many of which contained at least partial maps, created an environment in which many French people could experience their country in a new manner. The increasing ease of movement allowed many to better get to know the France that existed beyond their village. More travellers meant more map use. Maps allowed a first-time visitor to the capital to visualize the unfamiliar urban space before they stepped off a train. By purchasing and, hopefully, understanding a map before arrival, the visitor would recognize the layout of the city and feel more comfortable navigating its busy streets. In this way, increased map use further demonstrates how French society became more comfortable with its capital city.

The interaction between a map and its reader is more cryptic than in the case of guidebook commentary narrative, which can be quite explicit, with itineraries and suggestions for the uninitiated. Cartography lacks the fundamental components that help readers understand books. Cartographers replace text with symbols and the map legend serves as a table of contents. As with any successful communication circuit, messages provided by an author must be received and processed correctly by a recipient (Darnton, 'First Steps' 175). Without a sizeable portion of narrative to guide the reader, an unclear map risks confusing the recipient, which will render it useless. Even when examining traditional books, reception is often difficult to decipher correctly. Reader correspondence, censor reports, library catalogues, subscription lists, and the study of book advertisements have all contributed to a better understanding of book reception theory, but it remains an inexact science (Darnton, 'First Steps' 156–63). This shortcoming does not diminish the role of the mapped message but instead reaffirms the importance of the cartographer's need to transmit knowledge effectively. A successful map introduces and clarifies an unknown terrain to its reader. Although 'the linking of text and image [has been] frequent,' Barthes states that regardless of what text is included, 'a linguistic message is present in every image' ('Rhétorique' 49). Maps are an important combination of text and image where the cartographic narrative now plays a much more important role in delivering a message.

While map 'text' does exist in titles, legends, and tables, the uninitiated map reader generally only sees those components as part of the overall image. To amplify their significance, the cartographer has tools at his disposal, such as stylized text or colours and shapes, in order to stress key points. In this case, the text's appearance sometimes matters more than its semantic meaning. The use of bold type or capital letters, for example, will set the text apart and signal its importance. Therefore, it is necessary to examine both text and image, along with social and historical factors, to understand cartography and map discourse.

With this chapter, I hope to have illustrated both the challenges and that insights to be gained from the examination of historical maps. The success of the cartography published between 1889 and 1934 depended greatly on several factors. Access to education, a desire or need to understand a geographical area, or even the ability to travel throughout the nation all determined who saw maps and how that public reacted to them. In this book, cartography from Third Republic France focuses on Paris and demonstrates the changing idea of urban space. The Paris which Alphand presents in 1889 differs greatly from the Paris which Prost planned 45 years later. In addition to the amount of urban space portrayed, the documents also reflect the changing map-reading audience. Improved education, along with a uniform, understandable cartographic base from which they could create their new versions of the French capital, allowed men such as Jaussely and Prost to produce more specific cartographic products and to feel confident that readers would receive their message.

The maps in this book display the government's view of Paris and the surrounding region. Between 1889 and 1934, officials such as Alphand, Jaussely, and Prost showcased both the capital and its potential. In order for the government's vision to succeed, it would fall to republican politicians to change French society and create competent map readers. Over the course of the Third Republic, France transformed its population into a people that not only gained access to cartography and recognized the nation on maps, but also understand representations of smaller geographical areas, such as regions, departments, or (especially) cities, such as Paris. The confluence of factors discussed in Chapter 2 demonstrates how men such as Vidal de la Blache and Jules Simon succeeded in doing just that.

Chapter 2

Creating Map Readers

The Rise of Geography and Cartography in Nineteenth-Century France

> Notre enseignement supérieur ne tend à autre chose qu'à former un peuple de dilettantes et de beaux esprits, d'avocats diserts, de penseurs de salon et d'écrivains agréables. Aucune des connaissances modernes n'a pu franchir le seuil de nos Facultés: ni la géographie vraiment scientifique, ni l'économie politique, ni l'administration comparée, aucune enfin de ces sciences contemporaines, nées depuis un siècle au plus, et déjà adultes aujourd'hui. De là vient l'esprit superficiel qui, du sommet à la base, se répand sur toute la société française; de là viennent aussi cette légèreté et cette imprudence pratique qui nous ont précipités dans de si terrible désastres.
>
> —Paul Leroy-Beaulieu in *Le Journal des débats* (1871), qtd in Ozouf 25

After considerable retrospection, politicians and military officials came to believe that France had lost the disastrous Franco-Prussian war in part due to the inability of French soldiers and commanders to perform such simple tasks such as read a map.[1] Through providing insufficient support for education, industry, and science, they believed

1 In *Le Nationalisme français*, Raoul Giradet examines the rise of nationalism in late nineteenth-century France and its expression in political and academic discourses. He examines the work of leading thinkers, many of whom looked at the Franco-Prussian War as an indication of what did not work in French higher education and society. The German education system, for example, put greater emphasis on both science and geography.

that Napoleon III had not prepared the nation for the confrontation, and that the rapid, humiliating defeat caused by 'l'ignorance des officiers français qui auraient été incapables à lire une carte d'état-major' (Fierro, *Dictionnaire* 222).[2] Regaining the nation's lost prestige required a critical self-examination and a change in how it viewed and understood itself. One way to address these weaknesses was to improve and reform education. The current system's creation of 'dilettantes' and 'penseurs de salon' did little to help the military defeat of its neighbour rival, so France needed to incorporate more science into its national curriculum. As part of this drive, geography, once marginalized as a subset of history, became a vital component to improving national security and throughout the Third Republic the nation introduced the discipline to school children on an unprecedented scale. Once very rare, by the start of the First World War maps appeared more frequently in textbooks, travel guides, newspapers, and official government documents.

It is under these circumstances that France witnessed the birth of the French map reader, whose rise can be attributed to many factors. First, the average Frenchman became better educated. As education minister from 1879 to 1881 and again from 1882 to 1883, Jules Ferry instituted major educational reforms. In 1881, primary school became free. The next year, the government made it both secular and obligatory for children aged seven to 13. Within a generation, these pieces of legislation vastly increased the number of school-educated people in the Third Republic. As a complement to Ferry's educational reforms, the incorporation of geography, including the study of physical geography and maps into the child's curriculum via textbooks such as the *Le Tour de la France par deux enfants*, helped the French visualize and better know their own country.

In addition to showing children and their parents what France looked like in maps, the government could use geography in its promotion

2 Some historians even predicted that before 1870 any conflict between France and Germany would be hindered by the lack of geography in the French educational system. Vivien de Saint-Martin, 'bon cartographe et historien des explorations,' stated that 'la faiblesse des études géographiques est un triste symptôme dans l'éducation du peuple.' Others, however, were quite convinced that France could withstand any confrontation, regardless of the state of French geography and cartography. One Minister of War declared, sword in hand, that 'J'ai sur moi la meilleure des cartes' (Meynier 8).

of the nation and the Republic. The stinging defeat, combined with the annexation and attempt to 'Germanize' Alsace and Lorraine, pushed many French to rally to the republican cause. Maps introduced throughout the Third Republic served as visual proof that an historical region of France had been taken.[3] Furthermore, these and other cartographic examples made their way from the republican classroom into the minds of generations of schoolchildren who glanced up at the wall above their desk and, with little or no prompting from the teacher, could see a damaged France missing these two important regions. This proximity and access to maps also helped reinforce the message at home when an instructor or child discussed school work with parents. These constant reminders helped play to the republic's desire for revenge against her foe, but also provided a visual introduction of the nation to ensure that it corresponded with its overall concept of 'France.'

The private sector also played a role in creating map readers. The French became much more mobile throughout the nineteenth century. The large-scale expansion of the rail network opened up this mode of transportation to a greater sector of the population. That enlargement also introduced much of the country to citizens who, a generation or two earlier, may have never ventured beyond the confines of their village or department.[4] To facilitate this development, the travel industry became a significant cartographic force in its own right and those who travelled could use and expected to see city plans in guidebooks. By the start of the twentieth century, this tourism-based cartographic growth had eliminated the unknown in disparate parts of the country. The Côte d'Azur, for example, developed and thrived in part due to improved rail and road access, but also due to significant promotion from privately published guides, such as the *Guide Michelin* (Olson, 'Driving a New Tourist' 212). Conversely, guides also introduced Paris to the provinces. Many provincial French came to Paris for the city's many Expositions Universelles in the second half of the century. Those visitors often arrived with guidebooks

3 Vidal de la Blache's *France Départements* (1885) is one such example that was forced to shade Alsace-Lorraine a less bright colour than the remaining French departments.

4 For an 1850 map demonstrating the rapidly expanding rail network, see Duby, *Histoire de la civilisation française* (vol. 2, p. 243). The same network, greatly expanded by 1890, appears on p. 291.

devoted to Paris in hand, many of which contained specialized maps of the exhibition grounds. In addition to helping fair goers navigate the exhibition site, supplemental guidebook plans of other neighbourhoods provided readers with the means to navigate the capital, enjoy Paris's Haussmannian transformations, and take advantage of shopping and cultural options not found at home.

Outside of education and travel, other cartographic agencies of the time – for example the army's Dépôt de la Guerre – and city and government bureaus – such as the Service des Travaux de Paris – also published and sold maps to the public displaying their work and its influence on urban change, all of which helped promote small-scale map use in Third Republic society. Finally, France's colonial ambitions created markets for new maps. Cartography offered easy, safe glimpses of new exotic far-off lands as civil servants, businessmen, and travellers headed to various colonial outposts.

Geography's Appeal

Historically, map reading grew from a complex relationship between education, money, and power. The abstract thought required to understand cartographic works had been limited to those who had the means to devote time to education. Nevertheless, by the early years of the Third Republic, geography and cartography were exposed to a larger audience. The evolving map-reading public served as an indicator of how Third Republic map readers understood what they viewed. As maps became more commonplace, the French needed to be able to read and understand them, and the education system played an important role in helping create map readers. Before the geographer and historian Pierre Émile Levasseur[5] assumed the chair in Géographie économique et statistique at the Collège de France in 1872, geography occupied a very limited place in the French curriculum. Its only inclusion existed in courses devoted to the subject's relationship with history. Before him, 'la plupart de ceux qui s'étaient intéressés à

5 Selected by Jules Simon, the Education Minister in 1871, Levasseur and Alsatian-born Auguste Himly put together the *Rapport général sur l'enseignement de l'histoire et de la géographie*, investigating the inferiority of the French education system to the German (Claval 27).

la discipline, le rôle de celle-ci était de dresser le décoré dans lequel l'histoire s'était déroulée' (Claval 64). However, throughout the Third Republic, geographers such as Levasseur and Paul Vidal de la Blache (henceforth Vidal), transformed geography from a mere stage setting for important historical events into an esteemed university subject.[6]

Until the work of men such as Levasseur, Vidal, and later Jules Simon, the lack of geography instruction in schools meant poor map access and little or no comprehension for the average nineteenth-century French school child, most of whom remained unable to recognize even the most rudimentary cartographic materials. With no exposure to maps, the average Frenchman could not hope to understand them well enough to locate their home town or village, let alone make use of them as a soldier on the battlefield. Identifying the point at which the French public became map-literate is challenging. Geographer Christian Jacob theorizes the complexity in determining what he calls 'map literacy,' which 'implies networks of diffusion and social practices of readership' ('Toward a Cultural History' 192). Recognizing historic text reception has always been a challenge. At what point could the average French adult look at a map and confidently read it? More generally, how did a map's appearance influence its reception? The combination of steady exposure to the printed word and accompanying technology ultimately determines any reader reaction to a text (Darnton, 'What is the History of Books?' 135). In nineteenth-century France, the printing industry took advantage of technological advances to lower distribution costs and improve quality. New printing techniques gave publishers the ability to create colourful and detailed texts more uniformly. In addition, enhanced distribution allowed more French people to see them. A greater, more varied use of colour, for example, made maps brighter and more attractive. Lower production costs meant that publishers could place more maps in a larger variety of books and atlases.

These developments suggest that the French map market followed that of book publishing as a whole and responded to what people

6 For a better understanding of the history of geography in France, see Claval's *Histoire de la géographie française de 1870 à nos jours* and *Autour de Vidal de la Blache, La formation de l'École française de géographie*; Howard Andrews's 'The Early Life of Paul Vidal de la Blache and the Makings of Modern Geography'; and Vidal de la Blache's 'La Conception actuelle de l'enseignement de la géographie' and 'Leçon d'ouverture de cours de géographie' in his *Annales de Géographie.*

wanted to read or buy. Regarding map comprehension and use more specifically, Jacob asserts that

> a major need in the history of cartography is to comprehend the extent and the limits of maps in a society. Unlike other printed text, historic map use is probably the most difficult single aspect to ascertain, since when people look at maps they often leave no visible marks. Unlike book readers, who might annotate a text, their vision is invisible to us. (Jacob, 'Toward a Cultural History' 192)

In that respect, any determinant of map use within a given society or among a social class provides a better understanding of map-makers and publishing institutions overall. Jacob cites medieval *mapaemundi* as examples. Maps from the Middle Ages, many of which were employed not to give direction but to provide decoration, could not 'be understood without examining the intellectual practices of clerks and monks and the ways they dealt with written texts and images, the religious context of the particular culture and the contemporary interpretation of ancient encyclopaedic works' (Jacob, 'Toward a Cultural History' 193). Therefore, insight from preliminary knowledge acquired from the national education system during the Third Republic helped nurture informed French map readers appeared and will aid present-day researchers to recognize how maps may have been received.

As the maps discussed in this book appeared in print throughout the Third Republic, geographers in the French education system were building an audience for them. From the fall of the Second Empire to the First World War, the French publishing industry became more specialized, literacy spread, and a larger selection of scholarly textbooks appeared in classrooms. Yet even this expansion initially limited the amount of cartography seen by French children as geographers were forced to start from a very limited base. Only a few geographic *manuels de classe* and atlases published during the nineteenth century contained maps. Joseph Gilbrat's revised *Traité de la géographie moderne* (1813), for example, listed longitudes and latitudes and featured a list of major cities, but provided no maps. The nation's primary history and geography textbook, *Le Tour de la France par deux enfants*, contained no maps at all before 1884, and none of France as a whole before its 1905 edition (Weber, 'L'Hexagone' 1180). Therefore, it would take a few decades after the declaration of the Republic for all school-age children under the tutelage of capable teachers to have access to, and fully understand and appreciate, cartography.

Nineteenth-Century French Geography

Celebrated French geographer Vidal de la Blache perhaps did the most to introduce geography and, later on, maps into French classrooms. Educated under the Second Empire, Vidal lamented the traditional role of rote memorization in geographic training, which to his mind dulled the discipline and ensured that it served little purpose in the classroom. Even as late as 1905, 35 years into the Republic and 24 after the beginning of Ferry's reforms, Vidal complained that in France, 'la géographie est surtout une affaire de mémoire.' To make the discipline less reliant on lists of names and locations, he pushed for a more prominent role for maps in the national curriculum, stating that 'on ne peut s'empêcher de penser que les cartes peuvent servir [à l'élève] d'auxiliaires, et qu'il y a un meilleur usage à faire du temps des élèves que de leur enseigner de simples noms' ('La Conception actuelle' 194). If educators encouraged their students to explore the subject in abstract terms through cartography and shifted the classroom away from rote memorization, he believed that geography would become more interactive, interesting, and practical.

The Republic saw in the defeat of 1870 a sign of the need to transform not only its military, but French society as a whole. During the first half of the nineteenth century, instruction in France remained a privilege reserved primarily for children of the wealthy so that large sectors of the population remained outside the French educational system. Yet cost alone was not reason enough for the large absence of children in school. In rural communities, for example, many households did not have the luxury of parting with an essential farmhand during critical times of the year, such as the planting and harvest seasons. A few legislative advances attempted to change this custom and get students into the classroom. The 1833 Guizot Law created publicly financed *écoles des garçons* in communes with populations larger than 500 and during the short-lived Second Republic the 1850 Falloux Law, while granting the power of education to the Catholic Church, equalized education somewhat by creating *écoles des filles* in communities with populations greater than 800 (Weber, *Peasants* 35). The implementation of this system of free education in France dating from the July Monarchy was far from uniform, however, and few studied geography. History was the only subject that attempted to integrate it into its curriculum by casting the subject as a setting for historical events, and even then it was only introduced at higher levels, well after many students had already finished the minimum requirements. Geography at the primary school level, where it existed, focused largely on a few select items of vocabulary related to landscape.[7]

With the advent of the Third Republic, and the recognition that a lack of cartographic knowledge played a role in 1870, France sought to increase geography's presence in the school system. Yet there were challenges. The lack of geographers in French universities, for example, produced a system in which history teachers became the *de facto* geography instructors in schools. These individuals would then have to assume the front line in any new promotion and teaching of the subject. However, most history educators had received little or no formal training in the concepts and methods of geography and, lacking practical experience with the subject themselves, few instructors chose to include maps in their lesson plans (Martin, 'Le Monde des éditeurs' 191). The limited teaching of geography prior to the Third Republic meant that any use of the discipline depended solely on teachers' desire or ability. Furthermore, few administrators saw much value in the subject. One school inspector in charge of teaching geography in Second Empire France declared:

> On n'enseigne pas la géographie; on enseigne seulement la manière de l'apprendre. Composée uniquement de faits ou isolés les uns des autres, ou qui, du moins ne sont pas liés entre eux par un enchaînement qui existe dans d'autres sciences, la géographie est en grande partie le domaine exclusif de la mémoire: elle s'apprend par la lecture de l'histoire et des voyages. (Broc, 'L'Établissement de la géographie' 546)

7 Claval states that geography had a very limited role in French primary schools prior to the Third Republic: 'la géographie n'a pour but que de familiariser avec un vocabulaire descriptif de bas, la montagne, la plaine, le littoral, le fleuve, la ville, la forêt, l'île, la presqu'île, etc. et de mémoriser une nomenclature qui n'est pas toujours liée à l'apprentissage de la carte' (45).

Just as Vidal lamented, memorization – not map reading – was seen as the only proper way of disseminating and retaining geographic information and, since instructors were poorly trained, it was often easier to exclude maps from lessons completely. Although more prominent during the early years of a child's education, this cartographic gap existed and grew at every level. Even the Sorbonne, the apex of the French university system, which had included a chair of geography since 1809, continuously staffed it with historians, not geographers (Martin, 'Le Monde des éditeurs' 191). Further working against the academic subject, many school administrators had little regard for the discipline and felt that the French did not need to use maps. Officially, there was only one recognized reason for promoting geography under the Second Empire. Better knowledge of it could help the newly industrialized French business community sell goods abroad:

> l'enseignement nouveau était destiné à former des industriels et des commerçants. La géographie avait pour eux une importance plus grande encore que pour les élèves de l'enseignement classique et un intérêt immédiatement pratique: pour bien faire le commerce, il importe de connaître les pays dans lesquels le commerce se fait. (Broc, 'L'Établissement de la géographie' 547)

If it could not help business, then geography was best relegated to its role in history, a respected academic discipline.

After 1870, the Republic wanted to create active, engaged citizens and there was 'no better instrument of indoctrination and patriotic conditioning than French history and geography, especially history, which 'when properly taught [is] the only means of maintaining patriotism in the generations we are bringing up.' However, 'most teachers knew history badly, and geography even worse' (Weber, *Peasants* 333). As a visual component of this new republican ideology, 'maps [could serve as] didactic devises for the socialization of individuals, for the indoctrination of soldiers, settlers, or school children' (Jacob, 'Toward a Cultural History' 195). Of course, the teaching of these republican ideals via maps would prove useful only if they were understood. If a citizen could not comprehend a geographical image of France, then even more abstract notions such as citizenship or the new republican model of governing the nation would be difficult for a newly educated population to absorb. Prior to 1870, few poor Frenchmen had the means or free time to travel far beyond their home village and therefore may not have related to a neighbouring department, let alone other far-reaching regions of France that might not speak the same kind of French or respect the same customs.

The reference to a geometric form could help define the idea of France without all the complicated historical or cultural definitions. Seen as 'la creation d'un certain état d'esprit,' the Hexagon provided one such convincing visual image of republican France towards the end of the nineteenth century (Weber, 'L'Hexagone' 1171). Its simplicity, combined with its ability to incorporate mathematics, a separate academic field, further added to its appeal. Even today, this six-sided geometric shape remains a convenient symbol for defining historical and cultural metropolitan France. With easy-to-recognize visual references, including maps that provided a clear understanding of the shape and size of their country, republican citizens would be able to see, understand, and mourn French territorial loss on maps. If familiar with a 'complete' France, the loss of Alsace-Lorraine within a 'damaged' hexagon would be easier to identify and would hopefully provoke much-needed patriotic reactions from a larger percentage of the population.

A Geographic and Cartographic Tour de France

Levasseur, Vidal, and Simon had worked to bring geography to the French. However, the creation and direction of reforms from a bureau in Paris would need to eventually play out in local classrooms nationwide. Success there depended on the availability of people and materials. Augustine Fouillée's *Le Tour de la France par deux enfants*, published under the pseudonym G. Bruno, was the text that arguably did most to provide cartography and maps to the French outside the university or academic circles. Initially printed in 1877, the 1884 edition first introduced regional maps and those from 1905 on included a complete map of France. With each successive update, the text provided a simple cartographic initiation to millions of French school children. Fouillée's maps play a minor role in the book's narrative, but the fact she included them in a history textbook as early as 1884 indicates that she believed that school children, or at a minimum their teachers, would be able to see, understand, and use them as they learned about France. A 'manuel de géographie, précis de morale, livre de sciences naturelles' (Ozouf and Ozouf 291–92), *Le Tour de la France par deux enfants* tells the story of André and Julien, two orphaned brothers from Phalsbourg, a village in Lorraine who cross the Franco-German border under cover of darkness in search for a lost uncle. Starting out in autumn of 1871, André and Julien circled France along with the academic calendar and introduced the various corners of the nation to its readers. The text's narration promoted national unity and the French language, but the maps presented France and its regions to her readers. At the dawn of the twentieth century, almost every student knew the story – solidly inscribed into the French *mémoire nationale* – by heart (Ozouf and Ozouf 298). Fouillée's textbook, used until the Second World War, relied on a simple, easily digestible organization to promote unified French geography, history, and culture. Although Fouillée herself never attended university (Dupuy 130), her work became an effective tool in the republic's mission to create geographically informed citizens.

Fouillée's *Le Tour de la France par deux enfants* served as more than just a cartographic introduction. It also represents a crossroads between the old and new ways of understanding the nation. Throughout much of modern French history, the educated slowly began to see themselves as not only members of a local geographical location but part of the larger French nation. Literature provides a few key examples of authors using narrative as the primary method to 'map' France for readers, a process which demonstrates this method of identification. During the Renaissance, for instance, a narrative-based cartography played an important role in helping the French 'visualize' the kingdom. In Rabelais's *Gargantua*, perhaps one of the best examples, the protagonist's adolescence is 'defined by places visited,' and his education 'consists in tourism, and a nascent military way of "seeing the world" prior to committing himself to a site where the cosmos, like a book or a map, might unfold before him' (Conley 141–42). By describing the kingdom of France through text, Rabelais helped establish a common French identity, something that did not exist on a large scale in the Middle Ages. Furthermore, Rabelais provided readers with familiar place names and a common language, which combined to create a unified national character (Conley 144). In the sixteenth century, the French nobility followed François I as he toured the country. These voyages served as part of their education process and Rabelais recognized this shift. Through the creation of this mental map of

France (Tours was a certain number of days west of Paris, and Nantes was a certain number of days west of Tours), his public, the educated nobility and bourgeoisie, could better comprehend and perhaps even visualize the interior of the country. Conley points out that this mental map serves as an excellent demonstration of the first examples of the formation of the idea of France. Regarding Gargantua's travels, 'the user supplied an imagination of distances in studying the toponyms and the points of interest listed or graphed in books that were no longer than four-by-three inches' (Conley 144). Just as the nineteenth-century French would turn to a common national language as one definition of identity, culture, and nation, Gargantua's mental map served the same purpose. It provided the sixteenth-century literate population, which considered itself French, with a common territory to which it could relate, understand, and belong.

Fouillée follows the French template of narrative use to promote and define France. Her *Le Tour de la France par deux enfants* exemplifies this genre as it introduces the country to a larger segment of the population. Similar to Rabelais's Gargantua, who had travelled around sixteenth-century France, her textbook uses the boys' journey to supply a visual itinerary, one that all French children would follow and know by heart. By 1884, the author added cartography to complement her narration. With this one change, Fouillée democratized French cartography considerably. Not only did simple, basic maps appear for the first time in the hands of millions of school children; but those new readers could now hold, touch, and even copy or trace them to become amateur cartographers themselves.

Through narration, cartography, and the wanderings of two orphaned children from Alsace-Lorraine, *Le Tour de la France par deux enfants* promoted the idea that the many different regions of France were all part of one unified nation. Although much had changed in French society during the 350 years since the publication of Rabelais's *Gargantua*, French geographers of the early Third Republic remained reluctant to rely on abstract thought to teach and show France, preferring more concrete examples of Frenchness, such as rallying around a common history, goal (recapturing lost territory), or language: 'si on ne parle qu'une langue et que ce ne soit pas le français, alors quelle "honte"!' (Ozouf and Ozouf 282). Fouillée provided a cartographic model which would help students to think of the nation in broad, abstract terms. Her text further reinforces this ideal and facilitates the overall introduction to maps by making them not the primary focus of the text, but an accompanying reference tool to be used when needed.

The inclusion of maps in Fouillée's book illustrates the results of the efforts of French geographers, such as Vidal de la Blache and Jules Simon, to introduce the discipline to the French classroom. Her narration champions the republican values of honesty, cleanliness, and hard work but it also introduces a republican geographic discourse for the first time. In her textbook, the discipline further exemplifies Frenchness and Fouillée helps bridge the gap between an older, provincial vocabulary with the new push to republicanize the French curriculum, including geography. To do so, the author situates the towns and cities that the boys encounter within the actual departments but also keeps the old, outdated pre-revolutionary provincial names, suggesting that a significant portion of her readership still defaulted to the old classification. In terms of French geographic identification, Fouillée uses her cartographic discourse to reintroduce that republican administrative terminology. Her national and regional maps illustrate how France was divided administratively even if that cartography tended to downplay the republican nomenclature championed in the narration. Although outlined cartographically, none of Fouillée's map titles or legends acknowledge by name the then 86 departments in metropolitan France. This provincial-republican dichotomy exemplified the challenge faced by government officials as they pushed to create citizens, most of whom did not identify with far-off political titles as well as they did with the older, outdated provincial ones. Fouillée helps rectify this shortcoming with her narration, where she reinforces republican administrative jurisdiction by employing it heavily, but at the same time incorporates maps for the children to see and read.

Similar to Rabelais, *Le Tour de la France par deux enfants* relied on readers' ability and preference to follow narration. With their incorporation in 1884, however, Fouillée began to link regional maps with the story and the actual territory it described. School children could read about Bourgogne, find out where it is, and see what 'cette riche province' looked like in the book's map with the same name (Fouillée 104). As students tracked the boys' itinerary around France through both Fouillée's narrative and cartography, they also witnessed how meeting and interacting with fellow countrymen strengthens the nation. In a small village in the Dauphiné, for example, the two boys feel isolated, unable to communicate with their *hôtelière*, 'une bonne vieille [qui] ne comprenait que quelques phrases françaises, car elle parlait à l'ordinaire, comme beaucoup de vieilles gens du lieu, le patois du midi.' Moments later, a door opens and the hotelkeeper's children return from school. 'André, s'écria Julien, ces enfants doivent savoir le français puisqu'ils vont à l'école' (Fouillée 161–62). Language forms an instant bond and this situation proves the extent to which education unites all of France's children, one of the government's goals. Through a common language, Fouillée conveys to her readers that the Dauphiné region, like the Savoie preceding it, has unique characteristics. Its agriculture and industry nourished the area's economy but it represented just one part of a large complex, yet talented and integrated nation.

Geography and Cartography as Citizen Builders

Before Fouillée could publish her *Le Tour de la France par deux enfants* with maps, the cartography inside would need to be understood at a minimum by the *instituteur.* For this to happen, French educators needed to lay a solid geographical foundation. When Pierre Émile Levasseur assumed the chair in Géographie économique et statistique at the Collège de France in 1872, the subject occupied a very limited place in the national curriculum. However, in the first few decades of the Third Republic, prominent geographers such as Levasseur, and later Vidal, worked to expand the branch of study whose role was, 'pour la plupart de ceux qui s'étaient intéressés à la discipline avant [eux], [...] était de dresser le décor dans lequel l'histoire s'était déroulée' (Claval 64). Thanks to their efforts, by the turn of the century geography no longer simply complemented history, it emerged as a dynamic, functional discipline that could be used to promote France.

With very few geographically informed teachers or students in the classroom, there would have been little need or expectation for the original 1877 edition of *Le Tour de la France par deux enfants* to contain any cartography, yet Fouillée promoted maps and their use from the start. Before the boys can enter France, André and Julien's passage appears in peril when they lose their guide due to a fractured leg. However, André 'se souvint d'avoir vu dans la chambre du garde forestier une grande carte du département, pendue à la muraille: c'était une de ces belles cartes dessinées par l'état-major de l'armée française' (Fouillée 17). Fouillée shows how the self-proclaimed best student in

Phalsbourg not only recognizes the army's field map on the wall, but is able to study the countryside in order to find his way into France. In fact, he is so comfortable with this process that he even draws a rudimentary map of the route himself (Fouillée 17). Vidal hoped that teachers would eventually be able to do the same. Later in the story, as the boys approach Lyon, Monsieur Gertal, their travelling companion at the time, continues to encourage this practice. When Julien seems surprised by the size of the city, their guide reminds him that, 'Tu as une carte de France dans le livre qu'on t'a donné à Mâcon, [...] ouvre-le et regarde la position de Lyon sur ta carte!' Upon examination of the map, the boys realize that the city's location 'est magnifique pour le commerce d'une ville' (Fouillée 147–49). In an important step in his own geographical education, Julien is able to follow their route into Lyon on the map with his finger (Fouillée 148). Both of these experiences are far from the norm for the average 1877 school child, yet they suggest two new republican geographical ideals. First, they demonstrate how students (and Frenchmen) might use maps in everyday life. Second, they foresee a time when books with cartography would be a normally accepted and perhaps even expected reference tool.

Geographical familiarity is a great source of pride for the pair throughout their journey. Outside the village of Gex, Julien stops to absorb as much of the countryside as possible when it becomes clear that the trio will not have enough time to visit the Savoie region. The boy glances over at Lake Geneva and the Alps and promises to remember what the landscape looks like and do the same for other corners of the country. 'Comme cela, disait-il, je vais finir par savoir ma géographie de la France sur le bout du doigt. Quand je retournerai à l'école, je serai sûrement le premier, et je serai bien content' (Fouillée 94). Mastering this ever-important subject will place him ahead of his classmates, many of whom were only starting to discover what Julien experienced every day with his own eyes. Republican France serves him as his atlas and classroom; for the rest of the nation's school children, Fouillée's book will have to do.

Fouillée's cartography remains secondary to the emphasis on French republican ideals, yet she demonstrates the extent to which that philosophy played a role in Third Republic education. Hearing positive commentary about the new government in the classroom and looking to seek revenge on Germany, many students received the republican-trained and employed *instituteur*'s point of view of both France and the new regime. André and Julien, on the other hand, must learn these lessons from the citizens they meet along the way. Backed by republican-dominated ideology in texts like *Le Tour de la France par deux enfants*, schoolteachers were France's front line against religion, the monarchy, and foreign enemies. When the boys visit the capital, Fouillée devotes a full page to republican institutions such as the Corps Législatif, the Palais de Luxembourg, and the Élysée Palace. Yet her description of Versailles, perhaps the ultimate symbol of France's authoritative past, presents the Sun King's palace as nothing but a pleasant excursion from the capital. Other references to infrastructural embellishments made under past monarchical regimes, such as the Place des Vosges or the Pont Neuf are also missing. Even the Louvre is referenced only as a repository of art, not the royal palace it had been. As Fouillée silenced the monarchy in her text, the government was removing the Church from the classroom. Pro-government teachers hyped official messages of liberty, *laïcité*, and republicanism. These government missionaries spread out from Paris to local villages with the goal of solidifying the new French republican identity, uniformly educating its children, and eliminating social inequalities. Education was a common bond, a tool to better the nation's future and promote the government.

The discipline of French geography itself evolved considerably between Fouillée's inaugural 1877 edition, when she first sent her protagonists around the nation, and the publication of her 1905 version, which eventually included a full *carte de France*. An emphasis on wall maps, like the one André recognized and used, plus the other teaching aids and local field excursions at all levels of the programme were also important innovative elements of Simon's proposals (Andrews 178). Following these geographical curricular reforms and the maps they eventually produced, French school children soon began to recognize from a very early age that they were part of a larger community and no longer identified themselves primarily as residents of a local village. These future republican citizens now belonged to departments, to regions, to the French Republic and, ultimately, to the greater European continent. Few educators could argue that rote memorization of geographical terms and coordinates, the favoured method prior to Vidal's reforms, had been able to produce the same results. Furthermore, seeing maps and reading about children from disparate corners of the nation would also help students relate to their compatriots around France. School children in Paris or Lille may not have heard of the Puy-de-Dôme department or been able to locate it on a map before having read Fouillée's text, but they were more likely to recognize it once they read about Jean Joseph, 'le petit garçon de huit ans' from Auvergne who, like Julien and André, had no parents and liked to listen to stories of French history (Fouillée 131–32).

Thanks to this overall shift in geographical teaching methods, maps on classroom walls and in textbooks, such as Fouillée's, appeared with more frequency throughout the Third Republic.[8] Their employment remained inconsistent, however, and often depended on the competency of the *instituteur.* An untrained teacher might have little success in locating tangible local landmarks, such as a neighbouring village or a river that children might wish to identify on a large map of France. As Vidal's reforms came into place, many students were able to visualize their community cartographically for the first time. *Instituteurs* themselves became amateur cartographers in their own right, drawing and explaining maps on the classroom blackboard, just as André had done in order to prepare the route back into France (Fouillée 17). Thanks to professional and hand-drawn maps, students now saw themselves and France in relation to other places about which they may have read. No longer large and abstract, these relevant cartographic images soon became clear, concise tools which could be viewed and accessed daily. Furthermore, students began to locate their towns and villages correctly in relation to the rest of the country and Fouillée's text played an important part in eliminating this shortcoming. If a teacher drew poorly or had little knowledge of another region, he or she now had access to maps in *Le Tour de la France par deux enfants* or a walled *état-major* displayed in the classroom or school, both of which the children could reference, view up close, and even touch.

Teaching students about France through culture and history was the government's primary goal and maps, such as Fouillée's, helped in this endeavour as they allowed teachers to show their students what the nation looked like and where it was located in relation to its neighbours. Furthermore, politicians argued that cartography could bolster national unity as effectively as a common history. Identifying France on a map of Europe full of different nations and cultures could also instil pride and allow students to exclaim, 'That is France!' or 'This is my country!'

8 Vidal's series *Les Cartes murales* represents one prominent example of this trend. See Vidal's 1885 wall map, 'France Départements.'

Travelers had to understand guidebook city plans for them to be useful and classroom map reception faced the same obstacle. For students to recognize and use them regularly, maps of France had to be taught in a manner in which they could be understood. The embodiment of France through the Hexagon proved useful.

Over time, the use of maps in the classroom became even more effective as more students attended school and viewed them. The presence of more cartographic instruments (wall maps and textbooks) in schools made it easier 'de se référer à une image cartographique de la patrie' (Weber, 'L'Hexagone' 1179). Therefore, once students were able to comprehend a geographical image of the nation within Europe, then they might also be able to absorb even more abstract notions such as citizenship or a representative republican form of government. In many respects, Second Empire France had remained a country of regions and localities and any Third Republic textbook or classroom map would need to change that mentality, but only in a manner and pace acceptable to its readers.

French *instituteurs* helped republicanize their students through education, language, and now maps, and Fouillée did her part by referencing republican geographical titles at a time when many students might be able to name their own department, but few others. Throughout their adventure, the boys and their companions constantly discuss their physical location. This helps position the protagonists and illustrates how France's departments relate to its provinces. In addition, it provided another opportunity for the teacher to point out a location on one of Fouillée's maps. On that same visit to the small community of Gex, Fouillée states that the 'voyageurs, tout en causant, avaient depuis longtemps quitté le département du Jura; ils étaient maintenant en Bourgogne, dans le département de l'Ain' (93). Bourgogne, a historically important province, situates the perhaps less-well known Ain department and commune of Gex for the reader. Further along, when the boys arrive in the city of Bourges, André announces to his companions that it is 'l'ancienne capitale du Berry et le chef-lieu du Cher' (Fouillée 118). The central province helps position a potentially unfamiliar city and department for both teacher and student. This declaration highlights the fact that, although Berry itself no longer existed politically, it remained in the national geographic vocabulary. Yet, thanks to texts such as this, the populace would have the tools to switch and refer only to the Cher department in the future.

During the same conversation, Monsieur Gertal too bridges the two nomenclatures, stating that the neighbouring Bourbonnais, an old royal designation, 'a formé le département de l'Allier' (Fouillée 118). Examples like these demonstrate how Fouillée used her narration to situate a department or its *chef-lieu* within an old former province or even the larger nation. Given that France had 86 departments at the time, many named after far-off geographical elements such as a river or mountain chain, locating one without familiar reference points would have represented a challenge for the population as a whole. On the other hand, by linking a department to a region listed in her table of contents or found on her maps, Fouillée provided teachers with helpful reference points and made it easier for students to find relevant information.

The symbiotic relationship between provincial and departmental names dominates the book's geographic discourse. One notable exception to this practice is Avignon. During their discussion with André and Julien, the local French-speaking school children constantly refer to the area in pre-revolutionary terms: 'Vous connaissez bien le comtat Venaissin, qui est tout près d'ici? A cette époque, [il] appartenait aux papes' (Fouillée 164). Later, as the boys arrive in Avignon itself, the narrator too employs this terminology. From their train carriage window, they see 'l'ancien château où les papes résidaient lorsqu'ils habitaient le comtat Venaissin, enclavé dans la Province' (Fouillée 175). It appears that, even centuries later, the lingering vestiges of France's papal palace remain integral to the area's identity. It is not until further long in the boys' expedition, after they have travelled to the other side of the country and identified many more departments, that Fouillée finally corrects this variance. Julien's history textbook, to which he refers often, lists manufacturer and inventor Philippe de Girard's birthplace as a village in the Vaucluse department. Since this is the first time readers heard tell of that location, Julien helps out by declaring: 'Le département de Vaucluse, se dit Julien, chef-lieu Avignon' (Fouillée 261). By changing the administrative discourse at this point, Fouillée reminded readers that even though provincial terms were used in some regions more than others, she needed to rectify this anomaly and reiterate the fact that republican department names were the preferred nomenclature, even if her maps tell otherwise. Nonetheless, Fouillée's text served a greater purpose: introducing maps to millions of school children in a uniform way as every child saw the same maps every year.

Accessible *Cartes de France*

The overall newness of French cartography and geography in the curriculum suggests that there would have been little expectation of maps of France in school textbooks prior to Fouillée's 1884 edition. Once included, however, their primary role in *Le Tour de la France par deux enfants* was to enhance her storyline by illustrating the boys' path. As described, Fouillée's protagonists and narrator happily promote republican terminology throughout their journey, yet her cartography appears to minimize that nomenclature somewhat and defaults to the former provincial one. For example, she includes a provincial 'Carte du Nivernais, du Berry, du Bourbonnais et de la Marche,' not a departmental 'Carte du Cher, du Nièvre, de l'Indre, de l'Allier et de la Creuse' to show André and Julien's route as they travel around that central region.

Even though the boys mention the Cher and Allier departments multiple times in their conversations, Fouillée's map title fails to provide those republican names (Fouillée 116). France's Ancien Régime provincial designation officially disappeared in March 1790. Yet when Fouillée's 1905 map legend describes the geography of the area, we discover that it is not departments but provinces which are 'parfois couvertes de landes et de marécages' (Fouillée 116). On the map itself, the provinces receive the largest and darkest font and are the easiest to perceive. With effort, however, the reader can locate the five departments' smaller typescript. Since the reader's eye first targets the most noticeable objects on a map, Fouillée bestowed additional significance on the area's former royal names. Therefore, the smaller departmental titles appear less important and Fouillée's map suggests that her cartography's task here was to notice the boys' positions, not teach students the titles of French departments. Even though they do not appear in titles or captions, advanced learners can find the departmental names if they know where to look on each map.

This model continues throughout the remainder of the text. In fact, with the exception of the Île-de-France, an area 'qui a formé cinq départements' (Fouillée 272), none of Fouillée's map titles or their legends acknowledge – or even mention – the 86 departments in metropolitan France. A student examining the text and looking only at the images and maps would not find any of those divisions listed. Even her *carte de France* at the end of the book, which finally provided daily physical access to a national map, the one document on which

Carte du Nivernais, du Berry, du Bourbonnais et de la Marche. — Cès provinces sont parfois couvertes de landes et de marécages, comme dans le Berry. Le Nivernais et le Bourbonnais ont à la fois une agriculture et une industrie très actives: le Berry produit beaucoup de laines. Dans la Marche, se trouvent de petites villes industrieuses, comme Guéret et Aubusson, dont les tapis sont renommés.

Figure 1. Fouillée, Carte du Nivernais, du Berry, du Bourbonnais et de la Marche.

Fouillée provides a clear shape of the entire nation, labels the country by former regions (Fouillée 318). Faint outlines of departments can be seen on that map, but without the names only a well-informed reader would be able to identify many of them. To be fair, its size and scale limited what information Fouillée could include. An overly crowded cartographic presentation here might confuse new student geographers and too much text could overwhelm a reader, which could render the map useless. Instead, the map's political simplicity makes it much easier to understand and promotes the overall nation even if it is at the expense of individual departments. Fouillée also uses this national

Figure 2. Fouillée, Le tour de la France, itinéraire d'André et de Julien.

map to send a secondary message. The shaded Alsace-Lorraine on the eastern border adds an additional level of discourse. Halfway between a dark Germany and a white France, the light grey section questions the legitimacy of German jurisdiction over the two provinces and, in many respects, pushes the French border back to the Rhine River where most felt it belonged.

In addition to internal borders, *Le Tour de la France par deux enfants*' cartography could also help students understand all of France's international ones. During much of the Ancien Régime period, France often deferred to 'natural' boundaries such as oceans, mountains, or rivers to define its frontiers. In January 1793, during the Revolution, Georges

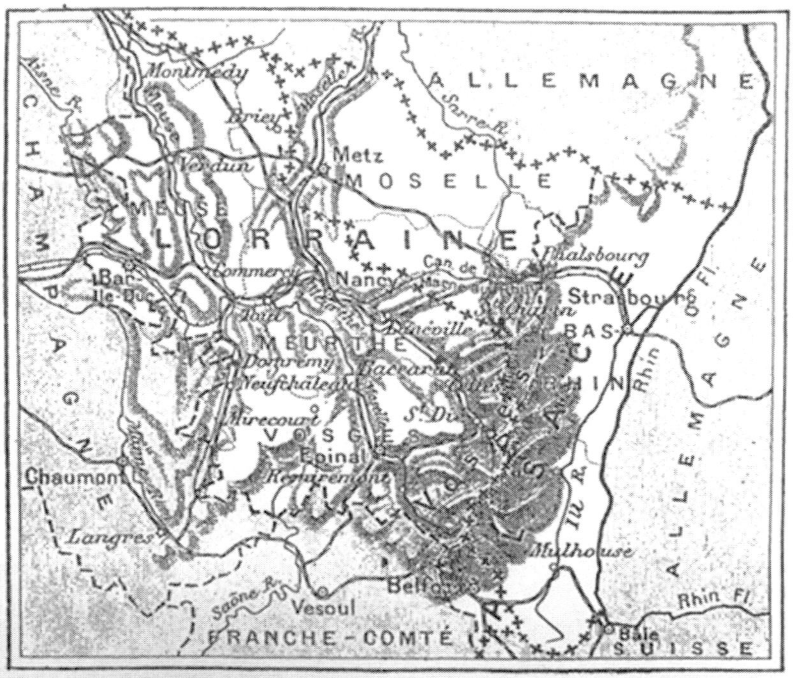

CARTE DE LA LORRAINE ET DE L'ALSACE, ET CHAÎNE DES VOSGES. — La Lorraine, séparée de l'Alsace par la chaîne des Vosges, est une contrée montueuse, riche en forêts, en lacs, en étangs et en mines de métaux et de sel. Elle a de beaux pâturages. Outre le blé et la vigne, on y cultive le lin, le chanvre, le houblon qui sert à faire la bière : l'agriculture y est, comme l'industrie, très perfectionnée. Une partie de la Lorraine et l'Alsace entière, sauf Belfort, ont été enlevées à la France par l'Allemagne en 1870.

Figure 3. Fouillée, Carte de la Lorraine et de l'Alsace, et chaîne des Vosges.

Jacques Danton invoked what he perceived to be the true limits of the nation as 'telles qu'elles sont fixées par la nature: Le Rhin, l'océan Atlantique, les Pyrénées, [et] les Alpes' (Weber, 'L'Hexagone' 1175), and most French concurred. Natural borders were simply much easier to justify and enforce than man-made ones. The creation of controversial political boundaries, such as the post-1871 French-German border, was challenging for a largely illiterate population to understand. Finding themselves on the wrong side of the new Franco-Prussian divide, Fouillée's two boys had first-hand experience of the importance of national sovereignty. At the beginning of the story, it is the army's *Carte d'état-major* that helps André and Julien identify that ever-important line of control. On the ground, however, the frontier was much less obvious. So the pair relied on stars and a few prominent landmarks, such as an old tower, to keep track of their progress. It was not until they passed the final mountain that André and Julien could be sure that they had arrived on French soil (Fouillée 25). With German soldiers on the lookout for people crossing the border, the physical ordeal was dangerous and Fouillée's readers witness the difficulty of trying to differentiate between nations solely on natural markers. Maps, such as Fouillée's 'Carte de Lorraine et de l'Alsace, et chaîne des Vosges,' however, documented the exact location of that divide (Fouillée 16). With this cartographic access and knowledge, students could see the practicality and ease of detecting the ambiguous border from their school desk rather than doing so in a dark forest.

This same issue arises later as the boys approach the Franco-Swiss border. As with André and Julien's border crossing at the beginning of their journey, M. Gertal too depends on physical markers to identify the limits of France: 'A nos pieds est la Suisse, mais à droite, c'est encore la France qui se continue, bornée par les Alpes' (Fouillée 86). Fouillée recognized the practicality of using natural barriers like mountains and lakes to identify the international border, yet also suggested a better way. If the reader turned to the 'Carte de la Franche-Comté,' they would see the border, clearly marked by a line (Fouillée 77). Students would not need to spend years exploring Lake Geneva on foot when cartographic science could identify the border for them. In practical terms, students learned the importance and role of national borders. Yet they soon understood the overall concept of sovereignty and how a line on a map represented much more than a simple mark on a piece of paper.

Vidal de la Blache and the Rise of French Geography

As the idea of man-made nations and borders entered the French educational consciousness, the lack of representations of France in atlases or geography manuals, many of which were published without maps, was of primary concern to those who wished to include geography in the national curriculum and use it to promote French ideals outside Fouillée's *Le Tour de la France par deux enfants*. If present in the national discussion, maps comprised just one of many rallying points for the

Republic. As the many obstacles to their wide-scale employment attest, the absence of cartographic images was not exclusively due to a lack of desire, but also of convenience. In the first two decades of the Republic, a bust of Marianne at the town *mairie,* for example, would have been more accessible to most villagers than a recently published topographic atlas or large expensive wall maps. But cartographic documents such as the army's *état-major* maps and Fouillée's *Le Tour de la France par deux enfants* began to change this situation.

While Fouillée's text was becoming part of the national curriculum, Paul Vidal de la Blache continued to promote geography. He was instrumental in reshaping its role in the Third Republic and his career mirrored the subject's rise under successive governments. For example, starting in 1885, Vidal's large, colourful *cartes murales* appeared in classrooms throughout the nation. Later, his 1903 *Tableau de la Géographie de la France* confirmed geography's newly found position in the French academic and scientific communities. In that work, which served as the introduction to Lavisse's *Histoire de France depuis les origines jusqu'à la Révolution,* Vidal turned to geography to conceptualize the idea of the nation, asking, 'comment un fragment de surface terrestre qui n'est ni péninsule ni île, et que la géographie physique ne saurait considérer proprement comme un tout, s'est-il élevé à l'état de contrée politique, et est-il devenu enfin une patrie?' (*Tableau* 8). Vidal tied land surface to identity and used geography and abstract thought to explain France's *raison d'être.* This justification was not surprising since others were also striving to define the French nation, whether through language, culture, or race. Maurice Barrès (1862–1923) turned to a common history and collective memory to define France. Charles Maurras (1868–1952) declared that 'la nation,' not the Republic 'est la plus haute des réalités politiques' (qtd in Giradet 198). Citing race, language, common culture, geography, religion, and military interest, Ernest Renan contrasts with his contemporaries. He contended that while all of the above were important and present in most 'nations,' his final definition is: 'une nation [est] une âme [et] un principe spirituel' (31). For Renan, a national identity cannot be demarcated solely on paper or defined in books, it must come from within the citizen.

If maps facilitated national identities along with cultural and historical sympathies, how could geographers get them into French schools in order for them to become part of the national narrative? Vidal's work on expanding the study of geography played an important part in increasing the discipline's reach. The son of Antoine Joseph Vidal-Lablache, a *régent de rhétorique* (teaching the senior language and literature class), he began his career in provincial France, teaching at high schools in Angers.[9] In 1872 Vidal was appointed to the chair of history and geography at the Université de Nancy. In 1877,[10] the same year Fouillée first published *Le Tour de la France par deux enfants*, he was appointed to teach geography at the École normale supérieure, a position second only to that of a chair in the Sorbonne in terms of prestige and influence, and higher than the Sorbonne in terms of its direct influence on the training of potential geographers among the nation's educated elite (Andrews 174). With his new post, Vidal's influence on the French national curriculum significantly enhanced geography's presence in the students' daily learning and instituted the move away from rote memorization to an increased use of visual tools such as maps.

Vidal was not alone in his tireless promotion of geography. Jules Simon, Minister of Education from 1870 to 1873, worked equally hard to encourage geography's rise. In September 1872, when he announced his intentions for curriculum reform in the secondary schools in a lengthy *circulaire* to school principals, Simon had already put in motion a variety of plans to expand and enhance the teaching of geography, modern languages, and physical education (Andrews 178). He had also formed a committee to look at the discipline's role in the overall school child's education. Simon's 1871 Commission de Réforme de l'Enseignement de la Géographie transformed French geography's position in French schools.[11] The government established the commission not only to revise its own curriculum, but also to provide clear advice on which books, atlases, maps, globes, etc. could be recommended for classroom use (Andrews 178). The work of Vidal and Simon paralleled the large continental shift in the discipline taking

9 Guiomar provides a detailed description of Vidal de la Blache's career, from his entrance in the École normale in 1863 to his career at the Sorbonne, which ended in 1908 (569).

10 Andrews points out that 'upward movement in a secondary school teaching career required a readiness to move from one teaching post to another.' As such, Vidal's youth was spent moving around France as his father slowly advanced in the *lycée* system (175).

11 For a list of members of the Simon Commission, see *Bulletin de la Société de Géographie*, sixth series, 3 (January–June 1872), p. 73.

place beyond France's borders. Four years later, the 1875 international geographic Congrès de Paris set forth goals and guidelines for enhancing geography education throughout Europe. These objectives included the study of topography at all levels of instruction, a better alignment between history and geography, the training of geography professors, and the creation of an international body devoted to geography education (Broc, 'L'Etablissement' 554). Instead of beginning the subject with the planetary system and having French school children work their way, during subsequent years, to France, the intent of the proposed programme was to turn this process on its head, beginning in the youngest years with a study of the immediate vicinity of the child's home, school, and commune, and enlarging the scope of enquiry and scale of material in tandem with the child's age and development. An emphasis on wall maps, the use of other teaching aids, and local field excursions at all levels of the programme were also important innovative elements of the Simon proposals (Andrews 178). As a result of these efforts, French school children now began to recognize from a very young age that they were part of a larger community. Outside of the classroom, other maps also helped in the promotion of the subject. The army's Dépôt de la Guerre sold departmental *état-major* maps to the general public as an informational tool but they often also served as wall decorations, as witnessed in *Le Tour de la France par deux enfants*. These easy-to-obtain map sheets cost much less than entire books or atlases, and played their own role in French geographical education, as demonstrated by André and Julien.[12]

The Simon Commission encouraged putting up wall maps in schools and, as a result, *carte murales* appeared in front of students with more frequency. Initially a large map of Europe or a French region had not been effective when one tried to locate tangible local landmarks, such as a village or a nearby river that children might recognize. Nevertheless, thanks to Simon, many school inspectors reported greater exposure to maps, which enhanced students' ability to recognize locations: 'Si les cartes murales et les globes sont loin de figurer dans toutes les écoles, les inspecteurs ont remarqué que les maîtres faisaient fréquemment usage du tableau noir pour dessiner le plan de la classe, du village, ou de la commune' (Broc, 'L'Etablissement' 548). Those

12 The *Melun Type 1889* sector (no. 65) of the *Carte d'état-major*, one such example, had a list price of 30 *centimes*.

maps allowed many students to visualize their classroom or community cartographically for the first time. Akin to De Certeau's view from above Manhattan, wall maps provided the first glimpses of large swaths of territory which could help explain to the French how their village related to surrounding areas.[13] No longer overwhelming and abstract, maps became clear, relevant, concise tools that could be viewed and accessed daily. Furthermore, once those rudimentary blackboard maps drawn by the *instituteur* became part of the school child's imagination, it is not hard to imagine that, 20 years later, the same students might better comprehend a map of a *département*, or a map of a never-visited, faraway city, like Paris.

Thus, just as the *instituteurs* used hand-drawn maps on blackboards to illustrate and clarify what was recognizable to villagers, the geographer used maps to make known the unfamiliar more generally. The act of looking at maps of larger areas (such as the nation, or the world) involves the viewer in paradoxical ways. It implicates at the same time a physical place (here or there), one's identity (how do I relate to this place?), and a reality (this place exists). Confronting a mapped world has intellectual consequences and is a major step in the process towards abstract thought (Jacob, 'Towards a Cultural History' 194). With more maps at their disposal, the Third Republican students' vision of France was no longer limited to understanding what they had memorized or what they had experienced locally first-hand. Maps provided pupils tools with which they were able to visualize themselves and their community within the larger French nation. Although still a complement to history at this point, geography slowly became an integral component in defining the French nation.

As Paris sent official government envoys, such as the *préfet*, into communities and departments to represent it, *instituteurs* also assumed the role of agents with their own promotion of the Republic. Daily face-to-face interaction with children helped them form a familiar bond and allowed instructors to use the curriculum to promote the Republic. Towards the end of the century, through his writings in his *Annales de Géographie*, Vidal facilitated geography's role in France's mission of creating citizens. Acknowledging the importance of a common history to the national identity, he reaffirmed that 'l'histoire dans les vieux pays

13 For a discussion of De Certeau's view of Manhattan, see Chapter 1.

que nous habitons, ouvre des perspectives lointaines; elle contient un riche patrimoine d'expériences; elle est mise, dès le premier âge, par des récits, en rapport avec l'esprit des enfants' ('Conception actuelle' 205). Vidal also began to espouse an idea of the larger geographical region based on his work in physical geography. In 'Des divisions fondamentales du sol français' (1888), he explored how some regions were formed naturally due to factors such as glaciers or mountains. For example, his definition of the Parisian Basin took into account geologic formations, which he then interpreted:

> Les relations des habitants n'y rencontrent pas d'obstacles. Elles ont trouvé, au contraire, un stimulant et une aide dans la disposition du réseau fluvial et dans la variété naturelle des produits [...] Le rapprochement des populations dans une contrée assez vaste pour embrasser le quart de la France, librement ouverte vers l'extérieur, a imprimé aux événements qui s'y sont accomplis un caractère de généralité qu'ils ne pouvaient avoir en aucune autre région de notre pays. Les différences locales s'y sont amorties. Les patois, en général, disparus. L'unité française y a grandi comme une plante naturelle. (Vidal 105)

Just as the root system of a thriving plant provides nourishment and a foundation, for Vidal the Parisian basin anchored and supported the overall rural, unified Third Republic.

Based in Nancy in eastern France, Vidal was also heavily influenced by the treatment of geography in central and northern Europe, primarily Germany. By the mid-nineteenth century French geographers began following the German example of promoting the more natural geographical divisions where French intellectuals had traditionally preferred dividing the country politically.[14] This *méthode naturelle*, using German practices as a model,

14 In 'Le Centre et la péripherie,' Maurice Agulhon focuses on the decentralization of the new political divisions. First Republic France took much administrative power from (royal) Paris and divided it up between the local authorities: '1789 renforce l'unification administrative et politique en remplaçant les provinces et autres institutions particulières par des départements, portions plus petites découpées dans le territoire national pour des raisons de commodité, administrées d'après les mêmes lois et règlements, et désignées par les noms tirés de la géographie physique objective (au lieu des noms de caractère historique, réputés entachés de "féodalité")' (827).

> est portée à perfection par deux Allemands, Gatterer qui divise le globe en grandes régions naturelles (Gebiete) inspirées des bassins hydrographiques de Bauche, et son disciple Zeune, inventeur des grandes divisions de l'Europe: terres carpatiques, péninsule balkanique, péninsule ibérique [...] Aux partisans de cette Reine Géographique où Géographie pure s'opposent les tenants de l'école politico-statistique, comme Büsching ou Achenwald, qui ne connaissent que les divisions politiques et administratives. (Broc, 'Quelques débats' 37)

This novel way of dividing Europe had a profound impact on French people's traditional views of themselves. Although most departments are named after rivers or mountains, their creation after the Revolution was primarily a matter of convenience, not any particular affinity for mountains, plains, or valleys. Originally 'un découpage purement géométrique,' the actual details adopted in 1789 decreed that 'la taille de tout département doit permettre à chaque citoyen "d'arriver de tous les points de ce territoire au centre de l'administration, en une journée de voyage," ce qui fond l'hypothèse d'environ quatre-vingts départements' (Roncayolo, 'Le Département' 2940–41). The Revolution had eliminated the large, historically prominent provinces to which the French still referred colloquially, as evidenced by their employment in *Le Tour de la France par deux enfants.* At the same time, it also created abstract departmental borders on a local scale for practicality, which, due to proximity and familiarity to residents, locals could better understand. Historically, kingdoms or empires employed natural divisions such as rivers and mountains to identify borders. By the Third Republic, geographers could promote the natural, geographic region in its own right, regardless of political or administrative agendas or practicalities. French geographers thus joined their European counterparts in identifying natural physical regions (i.e. the Parisian Basin) even as they turned away from previously used natural borders.

Fouillée too relied on the relationship between land surface and identity, and her work complemented that of Vidal, who had suggested that people could learn to rationalize France's *raison d'être* through geography. Neither exclusively cultural nor natural, for him France was

> l'abondance des 'biens de la terre,' suivant l'expression chère aux vieilles gens, qui pour eux s'identifie avec le nom [la

> France]. L'Allemagne représente surtout pour l'Allemand une idée ethnique. Ce que le Français distingue dans la France, comme le prouvent ses regrets quand il s'éloigne, c'est la bonté du sol, le plaisir d'y vivre. (Vidal, *Tableau* 50)

At a time when Maurras and Renan published their own definitions of a French nation, Vidal debated the idea that 'une contrée est un réservoir dont la nature a déposé le germe, mais dont l'emploi dépend de l'homme' (Weber, 'L'Hexagone' 1076). Although these definitions were varied and emotive, they helped solidify geography's place in the ongoing national debate over French identity.

Vidal differentiated between the physical region, often based on fluvial basins, and the historical one; and geography became a basis for understanding how administrative and natural divisions, although separate, complemented one another, and what influence they might have on political constructions (Claval 106). In the same manner, Fouillée attributed Lyon's commercial success to its access and proximity to rivers and rail lines: 'Situé à la fois sur la Saône et sur la Rhône, six lignes de chemins de fer aboutissent à Lyon et ses deux grandes gares sont sans cesse chargées de marchandises' (Fouillée 146). The Nord department too throve under the Third Republic in part due to its location, but even more so when combined with its industrious population and fertile soil. As the trio prepares to leave Dunkerque, their uncle tells André and Julien that 'C'est après le département de la Seine, le plus peuplé de France, et l'agriculture comme l'industrie y est prospère' (Fouillée 258). The republican political creations, such as Le Nord or Le Rhône, interconnect perfectly with the local physical geography championed by Vidal. Fouillée's school textbook provides capable and curious students the tools to find and learn more about each of them.

Fouillée's geographical descriptions, now in the hands of teachers, helped introduce the subject to school children, but only at primary level. In order to disseminate the subject more broadly, the system needed to help create future geography teachers for older students. Bringing innovative ideas as well as geographical basics into the French educational system required greater integration of the subject throughout the curriculum. To comply with the new European recommendations, Vidal needed to include physical geography in France's teacher formation, but there was no manual for this. When put in charge of the geography students at the Fontenay École normale supérieure, Vidal took it upon himself to author and publish *La Terre. Géographie physique et économique. Histoire sommaire des découvertes* (1883). However, he found it insufficient for properly training instructors of geography. So between 1891 and 1897, in collaboration with a former student, Paul Camena d'Almeida, he produced a four-volume *Cours de géographie à l'usage du secondaire* (Claval 63). These manuals established a uniform programme of study on a natural scale, which would lay the foundations for the discipline's place in school. Thanks to Vidal's efforts, qualified instructors would now be able to expand student numbers, teach the discipline correctly, and enhance the subject at a time when the government looked to expand its administration and oversight of new colonial possessions.

Travel and Colonial Geography

Fouillée's *Le Tour de la France par deux enfants* introduced geography and maps to French school children and Vidal helped train instructors to teach them how to read them. Beyond these two audiences, a small subset of the French population had been viewing and using maps for decades. For example, Second Empire rail expansion played a part in growing the French map-reading audience, though on a much smaller scale than experienced in national classrooms. Rail fostered the rise of the *bibliothèques des gares*, bookshops inside train stations that sold travel guides, many of which eventually included maps of rail companies' itineraries.[15] This field of texts became a major catalyst for private French map publication in the nineteenth century. Publishers such as Hachette and Chaix used rail bookshops to promote their own published *itinéraires* and travellers purchased them for access to detailed information on towns, which often included expanded narrative with descriptions of what to see, how to move about the city, etc. Most *intiéraires* strived to include every stop as part of their original mission since doing so broadened market share and accommodated the largest number of readers possible. For the larger, more frequented

15 For another source dealing with Hachette's growing rail station bookstores, see DeMarco's *Reading and Riding*.

cities and towns on each rail line, the guidebook urban hierarchy was self-evident: the more sizeable or important a community, the more narrative and cartography it received. The increasing reach of the railways as the century progressed meant the construction of new train stations and, eventually, additional station bookstores. As more people from provincial France moved to or visited large urban centres, these detailed city maps helped them get to know the ever-changing metropolises.

The French bourgeoisie took advantage of rail growth in the second half of the nineteenth century to travel around the nation. Imperial ambitions outside the Hexagon further opened up new territories to discover. France's expanding colonial empire created an additional market for maps, one that would cater to travellers, administrators, businessmen, and new residents, most of whom may have never set foot in the new colonial outposts. Between 1870 and 1878, leaders such as Léon Gambetta, Charles de Freycinet, and Jules Ferry promoted France's colonial expansion by using the argument that, for political and economic reasons, France needed to assert itself overseas. New colonies would help compensate for the loss of Alsace-Lorraine, as well as combatting French 'decadence' in the wake of Sedan and the Paris Commune and allowing the Republic to engage in the politics of grandeur. The government also made it clear that colonies were desirable because they would, as one member said, 'open up new fields for [France's] civilizing mission' (Conklin 12–13).[16] These endeavours helped distract the nation from its recent military and territorial loss, but also introduced the French to new parts of the world. Following the 1885 Berlin Conference, which reinvigorated European interest in the African continent, the French occupied and governed new territories in West and North Africa. To document French mastery of these large expanses of land, they needed maps. The Ministère des Colonies called for published maps and atlases of surveyed terrain,[17] which would allow French citizens to live, work, and travel in these territories.

16 For a more complete picture of debates surrounding French (and European) colonialism, see Conklin, *Mission to Civilize*; William Cohen, *Rulers of Empire: The French Colonial Service in Africa*; Driver, *Geography Militant*.

17 Its 1900 *Atlas des Colonies françaises* was proudly displayed at the 1900 Exposition Universelle (de Margerie *Exposition* 306).

France established its first major presence in North Africa with Algeria in 1830 under Charles X.[18] By the 1880s, Jules Ferry, the fervent proponent of the education reforms that eventually brought geography into the national conversation, proposed expanding French occupations in North Africa and Asia (including Indochina), though not every lawmaker endorsed Third Republic justification of colonization. French intellectual and political society was divided into those who believed that, as a great nation, France needed to 'civilize' people of Africa and Asia by introducing them to French values and culture, and those who felt that France needed to avoid such distractions and to concentrate on what went wrong and caused her defeat by Germany in 1871. Just as Charles X had hoped to divert an unhappy domestic population's attention when he sent troops to Algiers towards the end of his reign, the French government believed that an active international policy could help return France to its former grandeur and, in agreement with other European nations, decided to annex large portions of western Africa and Southeast Asia. Many politicians viewed the act positively as France's new colonial subjects would gain access to French language and customs and exposure to republican ideals. French colonial policy had for most of the nineteenth century aimed to assimilate colonial subjects into the French nation, and for the republicans, assimilation meant turning colonial subjects into republican citizens (Lehning 129). There was much debate about which peoples of the new colonies would ultimately be granted citizenship in the French Republic, and to what extent the government needed to map, and thereby control, these subjects and territories. With Ferry's reforms, the Third Republic had set in motion legislation to educate its own domestic population to help them become active citizens. Now successive French governments felt they could do the same for its colonies. The indigenous populations there needed not only to be 'civilized,' but

18 In 1885, Jules Ferry gave an impassioned plea for French intervention in Madagascar, citing many reasons, such as an increased market for French goods at a time when Germany and the United States were allowing fewer foreign goods into their markets; the humanitarian and civilizing aspects that the French Republic delivered to Algeria; and the maintenance of France's status as 'une grande nation,' keeping its position among 'le premier rang' instead of falling to the 'troisième ou quatrième rang' (Wieviorka 68–81).

also trained on how to become productive members of France or future leaders of its colonial possessions.[19]

Initially, the French knew little about many of these vast new territories, which would render any true occupation of them difficult. To rectify this situation, France could turn to its own recent past. The French had spent much of the nineteenth century documenting their national interior, a process that could serve as a model for doing the same in any overseas territory. In 1818, the army's Dépôt de la Guerre started surveying the nation for the *carte d'état-major.* These surveys covered both rural and urban areas. Prior to the Second Empire, authorities worried that Paris contained too many little-known, dark, dangerous neighbourhoods from which rebellion and revolution could erupt. To counter this threat, Napoleon III had Haussmann triangulate, record, and remodel much of the city's land surface in the hopes of better knowing and, by default, taming the disruptive capital. Although the *communards* proved the former emperor's idea to be flawed in 1871, the theory that knowledge equals control remained relevant. So, as parts of colonial Africa passed into French hands, France created several agencies to document the lands there. Le Bureau topographique de gouvernement général de l'Algérie and Le Service topographique du Régence de Tunis, both of which had ties to France's Service géographique de l'armée, sent engineers out into the field and acquired valuable information. Just as Haussmann had done to parts of the capital in the 1850s and 1860s and the Dépôt de la Guerre did for the rest of the nation, these agencies' subsequent surveying of the vast lands overseen by France not only provided knowledge to their administration, but also proved that the French could – and would – establish an effective presence on the African continent.

With reports and their data coming back from topographical agencies on the other side of the Mediterranean, French geographers

19 Paul Leroy-Beaulieu, a prominent liberal economist and moderate supporter of the Republic surveyed the history of colonialization in his *De la colonisation chez les peuples modernes* (1874). After 1880, he became an ardent supporter of imperial expansion. He thought that colonization was an activity limited to civilized peoples, and modelled the relationship between colonial power and colony on that between parent and child. He also suggested that colonizers would place 'the young society that it gave birth to' in a position to develop its abilities (Lehning 132).

at home could also promote mapping abroad. The creation in 1893 of the Chair of Colonial Geography at the Sorbonne helped facilitate this movement. Marcel Dubois, one of Vidal's initial students, first occupied this academic chair. A fervent believer in using geography to interpret and transform the world, Dubois believed that people outside the university, such as the military, sailors, businessmen, and diplomats, would benefit from a better comprehension of a colony's development potential, and that maps would provide it. An ardent nationalist, he also supported using geography to assist with subsequent French colonial expansion: 'La géographie doit être au service de la colonisation; elle permet d'orienter l'action de la France, de mettre en évidence les perspectives de développement et de comprendre les peuples indigènes. Elle doit aider à faire de la colonisation une œuvre civilisatrice' (*Géographie de la France et de ses colonies, Paris 1892*, qtd in Claval 80). In the final decades of the nineteenth century, the work done by the Dépôt de la Guerre, the Service géographique de l'armée, and their colonial sister agencies built up a sizeable corpus of reference documents of French administered lands at home and abroad.

This body of maps represented a source of pride for both French and colonial authorities. Multiple pavilions at the 1900 World's Fair at Paris prominently displayed these findings and used them to showcase French colonial territories. At the Algerian pavilion, for example, the Service géologique de l'Algérie displayed many maps. One of the most popular, its hand-coloured *Carte géologique de l'Algérie à 1:800 000*, documented the geological make-up of the colony and helped confirm French geographic knowledge and cartographic control of the sizeable North African territory. A wall in that same pavilion displayed five sheets of Louis Gentil's 1:200 000 *Carte géologique du bassin de la Tafna (Oran)*, which included the areas around the western administrative cities of Oran and Tlemcen 'Presque en entier,' as well as sections of Nemours, Lalla Marnia, and Sedbou (de Margerie, *Exposition* 308). These large, bright, and colourful maps drew in visitors and helped connect them visually with the French territory, even if many of the attendees would never set foot on African soil.

The Tunisian pavilion, another popular destination at that World's Fair, exhibited the Service topographique du Régence de Tunis's maps. The hand-coloured *La Carte forestière du Nord de la Tunisie* distinguished between different varieties of vegetation found there (de

Margerie, *Exposition* 308).[20] Maps of Afrique occidentale, Madagascar, Indo-Chine, and Nouvelle-Calédonie in their respective pavilions all presented far-off holdings and allowed French visitors to the fair to 'see' France's expanding colonial empire as well as providing the colonial authorities with the documentation necessary to solidify its control over much of Africa and Asia. Throughout the century, maps in schools, along with training and materials by Vidal and Fouillée, had introduced France to the French. Here, the medium helped explain to them in great visual detail the new far-off or exotic territories as well. In this capacity, cartography played its own integral role in the French 'civilizing mission' in extending the French ideal abroad.

Conclusion

One of the goals of this chapter was to recognize in context the role of geography and cartography in late nineteenth-century France. Cartographic accessibility and availability expanded greatly during this time, which allowed map readers to understand what they saw. This chapter explores the rise of what constitutes the Third Republic map-reading public. Fouillée's *Le Tour de la France par deux enfants* perhaps best exemplifies how one French text could play an influential role in disseminating cartography to millions of school children. Yet, as students read about André and Julien's voyage, Vidal de la Blache, influenced by academic movements elsewhere on the continent, fostered scholarship on physical geography. Better knowledge of the national terrain would assist the military in defending its borders, an aspiration which gained priority especially after the stinging defeat to the more cartographically proficient Germans. Furthermore, a greater understanding of France beyond the traditionally employed historical narrative provided further definitions and justifications of the concept of a nation at a time when that debate circulated throughout Europe. Finally, Vidal's publication of manuals to train instructors in the subject allowed the French to extend their geographic and cartographic literacy further. With this added instruction, more students would learn how to recognize and to read maps as the Third Republic progressed.

20 The map distinguished between 'les boisements de chênes-lièges, les chêneszéens [et] les broussailles' (308).

I am not suggesting that a simple analysis of the French educational system or cartographic technical advances is sufficient to understanding how Belle Époque society viewed and understood maps. However, better comprehension of the growing nineteenth-century French map-reading public aids our knowledge of how the cartography studied in the forthcoming chapters was both consumed and recognized, and it also influences how the government agencies could present the capital city to the French. Nevertheless, the information in this chapter offers an explanation of the role played by maps published during the nineteenth century in the official discourse of the Third Republic, what the French map reader saw, and, perhaps more importantly, how that reader might have reacted (if at all) to a published map.

For Paris, the largest and, arguably, the most important of all French cities, Haussmann's renovations, such as the wide boulevards and large parks, featured prominently in travel guide maps and could be seen as an endorsement of the previous regime. The arrival of the Third Republic meant that its governments needed to differentiate themselves from Napoleon III's regime, but also demonstrate their own investment in the nation and its capital city. With educational reforms in place and geographical and cartographic literacy on the rise, the Republic could employ maps, such as those discussed in this book, in its own citizen-building narrative. The *Travaux de Paris*'s map 'Paris en 1889: les opérations de voirie exécutées de 1871 à 1889' (discussed in Chapter 3), commissioned for that year's World's Fair, demonstrated the amount of work completed in the city since the fall of the Second Empire and the Commune. Furthermore, the Dépôt de la Guerre's decades-long *Carte d'état-major* project (discussed in Chapter 4) validated the regime's extensive knowledge of French territory and its ability to map it topographically.

The nineteenth century greatly changed French government, society, its population, and capital city. As a result of educational, social, and industrial changes, cartographic representations of France and Paris emerged with more frequency and larger percentages of the population could read and understand this expanded corpus of maps. This shift is noteworthy as we explore the cartographic representations of Paris throughout the Third Republic. Fouillée's *Le Tour de la France par deux enfants* demonstrated the benefits of a common language and history to its readers. If the author added a shared physical space, such as a local department or the Hexagon, that too would serve as an additional component to the child's budding republican identity. The fact that generations of school children viewed Fouillée's cartography meant that her maps helped form a common bond among them. Although her maps initially presented an outdated recognition of France's provincial regions, Fouillée subtly championed the current regime as she connected the old designations on the maps with the newer republican nomenclature in her narrative, making them more practical and useful. In the decades to come, the government, national agencies, cities, and individuals would publish maps and plans of cities, departments, and regions in the knowledge that the French educational system and a growing geographic and cartographic field of experts had formed a public able to read and understand what they saw.

Chapter 3

The Triumphant Republic

'Paris en 1889: les opérations de voirie exécutées entre 1871 et 1889'

> Ne devions-nous pas, à l'occasion du Centenaire de 1789, [...] glorifier hautement, cette République qui, aux jours de péril, avait ressaisi l'épée brisée de la France et avait su la rendre encore plus redoutable à l'adversaire, cette République qui avait ensuite cicatrisé les blessures de la patrie?
>
> —Jean Frollo (pseudonym), qtd in Wieviorka 144

The centennial of the storming of the Bastille prison in 1789 marked a significant milestone for the French republic. The subsequent hundred years had brought to France a return of the Bourbons, the Orleanists' July Monarchy, Napoleon Bonaparte and his nephew, and the Second and Third Republics. Unfortunately, most of these regime changes resulted from rebellion, revolution, or coup d'état. Needless to say, the chaotic nineteenth century did little to instil in the average Frenchman faith in the government's ability to remain in power for the long term. Yet, as the symbolic anniversary approached, there was a sense of optimism in the French capital for the last two decades had passed relatively violence free. Memories of the military defeat and the Commune remained fresh in many people's minds, but the ensuing two decades had confirmed the Republic's supremacy.

Even as the city prepared to celebrate, the Republic that existed in 1889 overcame many obstacles. The royalist Comte de Chambord's official rejection of the tricolour flag in 1871 effectively ended any hope of the Bourbons' return to power in France. Even as late as 1888, General Boulanger's political ascent briefly threatened the nascent democracy. In the city, the Third Republic experienced obstacles to taming Paris, as demonstrated by the anxiety preceding the funerals of Adolphe Thiers and Léon Gambetta. Early on, these events tested the government's ability to control Parisian crowds. Disruptions reminded the regime of the many revolts and revolutions throughout the nineteenth century and its perceived inability to govern Paris, and that major events could be either good or bad for the city. Even the selection, after much debate, of 14 July as the officially designated *fête nationale* in 1880 proved the difficulty of deciding how to recognize and acknowledge the Republic. Yet, as little as five years later, in 1885[1] Victor Hugo's trouble-free funeral procession and subsequent induction into the Panthéon would eventually demonstrate France's ability to manage its capital city. By carefully planning the route of Hugo's funeral cortège, the government was able to avoid problematic crowded areas of the city, and proclaim the city cooperative:

> In the end, the most direct route from the Arc [de Triomphe] to the Panthéon was chosen: down the Champs-Élysées, across the place de la Concorde and the pont de la Concorde, then left on the boulevard Saint-Germain to the boulevard Saint-Michel, the rue Soufflot, and the Panthéon. This route would not only avoid the east and the boulevards but take the procession past primarily governmental spaces: the Élysée, the Palais-Bourbon, and the ministries on the place de la Concorde. It also brought this republican ceremony through the heart of Paris, and staked out the left bank as republican territory. (Lehning 82)

The ability to control both Paris and its population allowed the Third Republic to spend its energies on bettering the nation and its capital city. In addition, the government could also use Paris to prove to the nation that the city stood solidly behind the Republic and its policies and that the city benefited from the new regime.

This chapter will examine Jean-Charles Adolphe Alphand's *Les Travaux de Paris, 1789–1889,* an atlas containing a series of maps

1 In 1878 and 1879 national celebrations took place on 30 June, a date with no historical significance. For those parliamentarians who wished not to commemorate crowd violence (the storming of the Bastille) in the national celebration, the date of 4 August, the anniversary of the end of feudalism, was also proposed (Lehning 65–66).

that present the spatial and technical organization of the city in the preceding one hundred years, published for the upcoming World's Fair. Through the 16 *planches*,[2] Alphand, himself Directeur des travaux, emphasized the continuity that he felt characterized the various transformations of the French capital since the Revolution. For this chapter, I focus specifically on *planche* 13, which presents 'Paris en 1889: les opérations de voiries exécutées entre 1871 et 1889.' It documents the government's ambitious Parisian road-building drive, in many ways a continuation of Haussmann's Second Empire campaign, though now through a republican lens. By planning and building these roads and then publishing a map certifying its progress, the government hoped to prove its commitment, at least on paper, to bettering all of the city's neighbourhoods. In addition, the work here conveys authority as Alphand provides dates of completion for specific roads. The 'Opérations de voiries' map celebrates the city and acknowledges the importance of an effective road network to Paris. A secondary drive behind this map plate was for the government to remind its citizens that the acquisition of knowledge of the entire city – acquired via road construction, surveying, and people on the ground – would eventually translate into effective jurisdiction and full control over the remaining rebellious or squalid neighbourhoods.

Alphand's 'Opérations de voirie' fits comfortably on one large sheet of paper folded over two pages. In fact, all his maps in *Les Travaux de Paris, 1789–1889* have the same scale and physical size, suggesting that, in many respects, any work accomplished after 1871 equalled that which had been finished under previous regimes. With the enlarged post-1860 city now in its third decade, the 1871–1889 map shows the reader to what extent the street works department had accomplished more than Haussmann. Alphand employed vivid colour

2 Alphand's 16 map plates are as follows: Plan de Paris en 1789 (*planche* I); Eaux et Égouts 1789 (II–III); 1789–1853 (IV–V); 1854–1878 (VI–VII); 1879–1889 (VIII–IX); Voie Publique – Reconstruction du plan de la Commission des artistes en 1793 (X); Plan de Paris en 1853, indiquant les opérations de voirie exécutées de 1789 à 1853 (XI); Plan de Paris en 1871, indiquant les opérations de voirie exécutées de 1854 à 1871 (XII); Plan de Paris en 1889, indiquant les opérations de voirie exécutées de 1871 à 1889 (XIII) – discussed in this chapter; Mode de revêtement des chaussées dans Paris en 1889 (XIV); Transports en commun – Itinéraires dans Paris en 1889 (XV); Édifices publics, Établissements universitaires et scolaires construits de 1871 à 1889 (XVI).

Figure 4. Alphand, Paris en 1889, Les Opérations de Voirie exécutées de 1871 à 1889.

to demonstrate the attention paid to the city and documents extensive completed infrastructure on all his plans. In fact, one might argue that his office has done too much. To the first-time viewer, 'Opérations de voirie' might appear to be a chaotic jumble of colours, lines, and symbols. His yellow, red, and black markings epitomize the Republic's presence but overwhelm the background and known landmarks.

Travaux de Paris transmits a message of progress to its audience, which fits well within the state's desire to promote a policy of improvement. Alphand happily showcased his agency's work in a positive light and correctly ascertained colour's ability to attract the reader's attention. The bright colours chart boulevards, streets, and other Third Republic infrastructure improvements, all of which could be considered part of a larger republicanization of the capital (including the construction of schools, police stations, and municipal buildings), a strategy the government wished to pursue at the national level. In many respects, the 'Opérations de voirie' *planche* helps the city conclude a relatively turbulent nineteenth century by presenting the public with proof of the Republic's achievements. Even though an initial reading of the preceding maps shows much progress in the Second Empire, in the 'Opérations de voirie' of 1889 Alphand happily appropriates much of Haussmann's work as his own. He prints the post-1871 dates of boulevard completion and launch, but does not indicate under which government the actual projects commenced. So, although Alphand credits all of this progress to the republic (it is on the 1871–1889 map), he offers a subtle nod to Haussmann, his former boss who initially planned much of the work.

From Empire to Republic

Map discourse has always contained secondary-level meanings, since cartographers use maps to portray their own view of the world. For example, prior to the 1789 Revolution, state maps in eighteenth-century France were published for both public consumption and state use (i.e. building a road or overseeing a battle), and cartographers often focused on promoting French strength. A common underlying goal of these documents was to use cartography to showcase the nation's power and influence, often at the expense of rival nations. So, map-makers presented France in a privileged way (via colour, map position, etc.). At that time, official French cartography often depended on private patronage beyond the sphere of the state in a nascent public, and 'social,' sphere, though that private patronage often linked back to the king or a royal minister (Joyce 39). With Napoleon's rise to power, French bureaucracy took a more prominent role in depicting France, whose portrayal became less overtly biased thanks to a more professional service, such as the Dépôt de la Guerre, as discussed in Chapter 4. Throughout the remainder of the nineteenth century, as the government transitioned from one run by aristocrats to a modern bureaucracy with universal male suffrage and education, the state took a larger role in producing its own national cartography. This transition took place at a time when the urban spaces to be mapped changed considerably.

As migration transformed peasants from rural France into workers in urban areas throughout the century, cities grew. Europe-wide, maps helped document the transformation of smaller, historic medieval towns and villages into large, teeming modern metropolises. Some of these metamorphoses happened from the top down. Picon notes that in French modernization,

> nineteenth-century politicians, engineers, and architects destroyed ancient cities while trying to preserve their essence. Haussmannian engineers and architects made frequent use of neoclassical compositional patterns supposed to be both efficient and picturesque. Among these engineers and architects, Alphand was especially conscious of this continuity between the techniques of *embellissement* and Haussamannization, hence his aborted attempt to use Verniquet's plan for *Les Travaux de Paris*. ('Nineteenth-Century' 144–45)

A benefit of the overall updating of Paris was to show off a final product, most notably around key events. During the Second Empire, Napoleon III held ceremonial inaugurations for the openings of boulevards. The Republic too needed to celebrate its achievements, and the 1878 and 1889 Expositions Universelles provided such an opportunity. As the opening of the 1889 Exposition Universelle approached, Paris had worked hard to repair and enhance its cityscape, yet there remained much work to do in creating and delivering a coherent, believable message to a large public that merged both the celebratory nature of the event with the stability and modernity that the Republic needed and desired.

The modern city of 1889, whose centre was now full of boulevards, showcased the best of France. Although introduced on a large scale under Napoleon III and Baron Haussmann, the Third Republic continued the trend of constructing wide avenues through the dense city. The excitement around another upcoming World's Fair only added to the desire to modernize the nation and build more avenues in outer *arrondissements*. However, following Haussmann's costly renovations in the 1850s and 1860s, the lack of finances at the beginning of the Third Republic reinforced a somewhat antagonistic view of provincial lawmakers towards Paris. City officials came to rely on these large, ceremonial exhibitions to advance their urban development agenda (Levin 22). Hosting the Exposition was one way to get new infrastructure at the government's expense. In 1878, for example, politicians wished to use that year's World's Fair to prove that France had recovered from the recent war and the Commune. Knowing that both dignitaries and visitors from around the world would travel to Paris to attend, city officials took advantage of the situation. They pressed the government for the necessary funds needed to complete the Avenue de l'Opéra,[3] the ceremonial link between the new Garnier opera house near the *grands boulevards* and the Louvre in central Paris. In their minds, the resulting beautification would impress and show visitors a repaired post-Commune city centre with completed boulevards, parks, and new exhibition spaces.

To truly modernize the capital in 1889, government leaders needed to do more than build fancy monuments or complete prestige boulevards. They would also have to address previously ignored *arrondissements*. If done correctly, the whole city would gain from hosting the upcoming World's Fair. As the press and notable figures focused on Eiffel's unpopular tower going up near the Champ-de-Mars, the government also needed to concentrate on the less sexy quotidian improvements that would improve the city's infrastructure and benefit the citizens who used it on a daily basis. The Eiffel Tower was a modern wonder that demonstrated French engineering skills, but it did not help people get to work or enhance trade with other countries. The government promoted the Exposition, but it also had to improve roads outside the

3 Although French streets and avenues traditionally use lowercase for *rue*, *avenue*, etc., I use capital letters in this chapter to mirror Alphand's capitalized nomenclature.

centre, expand the sewer system, build more parks, and open up much of the still-crowded city where the reputed denseness of Parisian slums stubbornly remained. Even though Haussmann drove the poor from much of the city centre, a general bourgeois 'phobie de la densité' of areas outside that zone remained (Charvet 240). This fear developed perhaps due to the perception of those non-documented and untransformed areas as 'unknown,' but also since the bourgeois rarely travelled to those unfamiliar, crowded neighbourhoods outside the centre. Accurate maps of a completed city would be one means to assuage such fears. The Commune, almost 20 years prior, remained a stark reminder to leaders that making little to no improvement to the dense working-class enclaves, predominantly in the eastern half of the city, would affect their desire to maintain their much-desired stability.

Beyond the eastern *arrondissements*, physical domination over the city would greatly help maintain social control. Throughout most of the nineteenth century, Paris's municipal council had prioritized bourgeois needs and wants, primarily at the expense of the working class. Jacques Lanquetin, a prominent wine merchant and proxy for the Paris city council, and one such example of this trend, examined social questions in the pre-Haussmannian city. Lanquetin argued most forcibly that the displacement of the Parisian population had already begun prior to the Second Empire and in 1841 published a series of papers full of statistical material to back up his case, which were widely reported in the press. His somewhat prophetic answer to modernizing and opening up Paris was to produce a coherent plan for the development of the whole city, specifically to transplant Les Halles to the left bank, where they would rejuvenate land values and general prosperity. Lanquetin's thinking here demonstrated a wider statistical insistence on concentration of populations as a key to economic progress and puissance (Green 47). It stood to reason that the deplorable conditions in crowded central residential blocks helped fuel dissatisfaction and dissent, so a less dense populace would certainly be happier and easier to manage. However, with limited funds, the government could not transform the entire city alone. Politicians hoped that instead of a fully financed set of works, publicly financed economic improvements in targeted sectors would help launch the remaining physical transformation, if those funds could be used to attract additional money and interest. The reconstruction of select neighbourhoods would hopefully also draw privately financed improvements, which would encourage more growth in adjacent areas.

Regardless of the government or city's intentions, in addition to the persistent economic segregation, social divisions remained. In the second half of the nineteenth century, and more particularly since the defeat of 1870, the French bourgeoisie was haunted by what it viewed as a social degeneration created in part by higher birth rates in lower-income households. In their minds, this demographic trend would lead to a lower quality population, as the poor would eventually outnumber the rich and reach a point where they could not be controlled, as had happened in 1871. This situation was further exasperated in Paris by the continued influx of additional workers drawn to the capital in search of factory jobs (Charvet 240). Equally impoverished as the workers already in the capital, these newcomers would further amplify the problem at hand as the city could not absorb many poor domestic immigrants without upsetting its current dynamic. This potential new urban environment frightened bourgeois residents, who pushed to limit growth, physically if necessary, as Haussmann did in decades prior. In many respects, the work started under the Second Empire, although linked to the previous regime, represented a solution for the current administration, as new wider roads would open up more of the city. Democratic laws now in place would, however, would make it much more difficult to justify and to enforce the same authoritarian policies that allowed Haussmann to transform so much of the city. Yet another large-scale urban transformation would allow bourgeois leaders to redesign and reformulate the city's infrastructure in areas untouched by the *préfet*, and to dictate how the working class would use it.

Haussmann's transformations were a source, product, and symbol of imperial policies for managing change. But now Paris, and through her the remainder of the country, would have to reconcile republican visions with the realities of a modern industrial society (Levin 21). As one can imagine, the republican government was hesitant to refer to Haussmann or give him too much credit for the way Paris looked. Anthony Sutcliff explains the high fiscal burden Haussmann's works placed on the finances of the Third Republic in *The Autumn of Central Paris*. Even though the city centre delighted residents and visitors, the election of a strong radical republican and socialist majority to the city council in 1874 seemed to suggest that Paris was ready to abandon the former *préfet*'s street improvement plan (44). But that position proved difficult to enforce. There were many compelling reasons to continue the major components of Haussmann's transformation plan. For one, the city had already started many of the *préfet*'s embellishment projects and, after 1871, they could not be stopped.[4] The destruction of large swaths of the city, along with pre-existing property sales, meant that, at least for the short term, the transformation of Paris would continue regardless of which regime was in control. In most cases, powers of compulsory acquisition had already been obtained by the developers, and it was in the city's interest to use them, for if they were to lapse, businessmen feared the national government might raise objections to renewal requests. Many opposed increasing debt, but Parisians wanted wide boulevards in their neighbourhoods (Sutcliffe, *Autumn* 44–45).[5] Even if they did not agree with the political and social ideologies of the Second Empire, the Third Republic's powers could not truly throw out its achievements; they would instead alter the character of modernity in the city to make it serve the middle class. Its leaders introduced their own political agenda focusing on science and technology, one that would liberalize and democratize the culture of change. Greater citizen involvement in government was achieved through universal (male) suffrage. At the same time, technological discoveries helped change how streets and buildings were constructed. Finally, a growing bourgeoisie called for greater stability. Renewed urban development in conjunction with international *expositions* fed a growing belief that progress was an evolutionary process towards republicanism – the apex being the 1889 Exposition Universelle (Levin 23). Yet, almost 20 years into the Third Republic, there remained much work to do. As a model republican city, transformed and modernized, parts of outer Paris remained unknown to the bourgeois elite who rarely ventured into the numerous unregulated slums and shantytowns on the city's edge. Ultimate success for the government would be to transform the whole city, to include more Haussmannian boulevards in more

4 Sutcliff writes that it 'would have been lunatic not to complete streets [of Paris] of which some sections were already in use' (*Autumn* 44).

5 Sutcliff proposes two additional reasons why the Third Republic needed to continue Haussmann's works despite the cost and inconvenience. First, foreign engineers, architects, and national governments were so impressed with Haussmann's achievements that they were planning similar improvements in their own countries. Second, no one had yet suggested a practical alternative to new streets as a means of updating obsolete urban areas to meet modern requirements (*Autumn* 43–44).

neighbourhoods, but sell the final product to the public as a republican triumph, not a former imperial one. Cartography was just one bureaucratic component of the Republic's mission to show its control and ability to govern.

Administrating Paris

Throughout the nineteenth century, scientific progress in the fields of health and hygiene combined with better access to information to improve the understanding of urban spaces, the people who inhibit them, and how administrators could regulate and improve them. At the same time, statistical knowledge regarding Paris appeared at a rapid pace. After the census of 1817, the government decided to publish data regarding Paris on a regular basis. In 1829, Louis Daubenton, Inspecteur général de la voirie, published his *Rapport relatif aux entreprises de construction à Paris et à l'interruption des travaux depuis cette année*, which detailed building activity in Paris. The inspector's report is still useful today for its wealth of data. It drew together three different forms of knowledge, whose combination helped structure the official portrayal of the city. First, Daubenton employed statistics to assess the capital in terms of quantifiable data.[6] Second, he investigated the state's role in the planning and regulation of urban development. Third, and perhaps most important, he incorporated new definitions of healthy and unhealthy space into his argument.[7] Taken together, these urban inquiries helped leaders develop a strategy for 'policing' the Parisian population in the broadest sense (Green 44). Later, similar studies also

6 Daubenton compiled numerous statistics on population. He showed 29,970 births in 1826, which meant that the city's population was actually 899,100 instead of the 890,431 the *Bulletin des lois* had originally stated (9). In a second example, he also looked at new construction. The 2,666 new houses built, combined with the 687 demolitions, resulted in an increase of 1,979. That number, added to the 26,801 recorded in 1817, left the city with a total of 28,780 habitations (qtd in Green 12).

7 For example, Daubenton pushed for wider streets for the creation of a better economic and healthy urban fabric. He noted that 'les propriétaires des maisons situées dans les rues étroites, et qui les feraient reconstruire à l'alignement, verraient le nombre de leurs locataires diminuer, mais le produit locatif de leurs maisons s'augmenter, parce que l'élargissement des rues y aurait attiré le commerce et l'industrie' (qtd in Green 72).

informed and educated leaders and scientists about the realities of city life. In 1840 Honoré Fregier, chief clerk at the Préfecture of the Seine published his *Des classes dangereuses de la population et des moyens de les rendre meilleurs*, which received the 'ouvrage récompensé' accolade from the Académie des Sciences Morales et Politiques. Fregier mobilized statistical data to show that, he believed, the capital held around 63,000 dangerous individuals. For him, that category ranged from people who did not work to those who would not work; from the inebriated to the depraved and criminal. Of course, this situation further confirmed the bourgeois stereotype of the poor. Harping on the usual environmentalist equations – laziness and drunkenness bred in conditions of overcrowding and squalor – it was hardly surprising that the author pinpointed the old decaying quarters as a font of vice. For the metropolitan audience, studies such as these reinforced the notion that the underclass was also the enemy within (Green 53). Begun in 1817, the Seine *départment* published eight volumes of a *Recueil des recherches statistiques sur la ville de Paris et le département de la Seine*. In 1865, this became a monthly bulletin and was soon supplemented by a weekly one. In addition to all the existing collections of numbers and figures, in 1880 officials decided to publish a statistical yearbook, the *Annuaire statistique de la ville de Paris* (Picon, 'Nineteenth-Century' 147). The publication, and sometimes manipulation, of data in official bulletins such as this lent further credibility to bourgeois leaders who viewed the populations of poorer sections of Paris as a danger that could easily spread. Politicians focused on the fact that the 'enemy within' had risen up once before, with disastrous results, and that it could not be allowed to happen again. In some respects, the increase in statistical data also mirrored the desire to know more about the city in non-quantifiable terms. Using various tomes published each year, one could count the number of people in certain occupations or how many Parisians lived in the sixth *arrondissement*, but if a dark narrow alleyway became closed off, or a new shantytown emerged, government oversight could easily miss it.

The demographic expansion in the capital between 1880 and 1910 quickly filled up most of the remaining land inside Thiers's wall and required a professional administrative branch of government. The personnel of the Agence des Travaux de Paris, many of whom had been assembled by Haussmann, provided that service to the nation. Adolphe Alphand took over after the *préfet*'s departure and helped continue to modernize the city. At 72 years old in 1889, Alphand represented continuity with the past and his experience ended up being a positive asset for the new regime. Even today, many politicians and historians refer to Alphand, the Inspecteur général des ponts et chaussées, Directeur des travaux de Paris, as the 'le grand 'aménageur' de son siècle' (Ory, *L'Expo* 14). In 1854, Haussmann brought Alphand, then Ingénieur en chef of the Gironde department, to Paris from Bordeaux. Once in the capital, Haussmann named him Chef des promenades et plantations and he directed the creation of parks and the installation of street furniture. Under his tenure, Alphand helped finish the Bois de Boulogne, but also had a hand in the creation and design of the Parc Monceau (1863), Boulevard Richard-Lenoir (1864), Avenue de l'Observatoire (1867), and the Parc des Buttes-Chaumont (1869). He was also involved in the design and construction of the Avenue du Bois de Boulogne (inaugurated as Avenue de l'Impératrice, now Avenue Foch) and the Parc de Montsouris. Named Directeur des Travaux de Paris in June 1871 at the very beginning of the Third Republic, he immediately combined the Services de la voie publique, des promenades de Paris with the Travaux du département de la Seine into one office. Seven years later, upon Eugène Belgrand's death in 1878, Alphand further consolidated and folded the Direction des eaux et égouts into his control (Rouvet 20). This final merger placed the city's sewer system under the same administrative umbrella as roads, which would facilitate development above and below ground in the outer *arrondissements*. The 5,000+ engineers now under his command, many of whom were trained under the Second Empire, became his new soldiers in the war on outdated infrastructure and his mission to modernize the city.

This series of grand mergers provided Alphand with increasing power and influence, which he used to change the city's composition. His work on the Parc des Buttes-Chaumont under Haussmann best represented what he wanted to do citywide with this consolidation of authority and symbolized the nineteenth-century ideal of using government to make the city a better place for residents. Repurposing this old dumping ground into a welcoming green space in Paris's north-east proved that science and engineers could change urban living for the better. Furthermore, the maps and diagrams of this work which Alphand published showed the technical nature of the transformation and even helped readers 'visit' the park before it was complete.

He was able to use that valuable experience to beautify other sectors of the city as well. Under the Third Republic, Alphand supervised the construction of 37 gardens and squares, besting the emperor, who completed 20 fewer. As Haussmann's protégé, Alphand had no qualms about using the office of the Travaux de Paris to further his cause and to demonstrate how much the Republic had done for the citizens of Paris.

Alphand believed that past regimes, although different one from the other, had pursued the same objective regarding Paris: to adapt it to the new needs of the post-revolutionary era. Via his atlas *Les Travaux de Paris, 1789–1889*, and through uniform graphic presentation, Alphand demonstrated continuity in the way this objective had been pursued by the various governments, a continuity he himself epitomized since he had served the Second Empire and Third Republic with equal dedication (Picon, 'Nineteenth-Century' 144). Attacked often by the left for his excessive powers and the lack of oversight of his engineers, Alphand evoked the best interests of the city with talent, diplomacy, and perspicacity. He believed that the exceptional power accorded to him immediately following the fall of the Commune in 1871 was justified due to the need to start the city's reconstruction quickly. He promptly completed both the Avenue de l'Opéra and Boulevard Saint-Germain. Before the Third Republic and prior to this 1889 atlas, Alphand was perhaps best known for *Les Promenades de Paris* (1867), a work that documented his vision for parks and urban space. In that tome, Alphand wanted to explore a history of gardens both public and private in Europe and elsewhere. His text explained how to merge the aesthetic and practical when adding green parks to the urban environment. He thought it important to 'préparer l'esprit du lecteur à la description technique de nos jardins urbains, de placer sous ses yeux un résumé de l'histoire de l'art des jardins, et d'indiquer ensuite les principes généreux d'une branche d'art, trop négligée de nos jours par les artistes.' By doing so, he could extol the virtues of additional public gardens and green space, the creation of which 'constitue une section importante de l'art général' (*Promenades* I). Thus, his Second Empire experience and aesthetic served the Third Republic capital well as he employed his vision of modernity in the Paris public works office.

Alphand used his bureaucratic army to transform the city. As a new wealth of information helped document the realities of nineteenth-century Paris city life, a new crop of administrators, many of them trained as engineers, entered government and began to leave their mark on the capital. The curriculum and philosophy at the École Royale des Ponts et Chaussées in Paris, perhaps the most prominent example of a state-run school that employed cartography and from which Haussmann drew many engineers and designers, formed an army of new disciples of science who focused on technical applications and design theory (Chevallier 38). Upon finishing their studies, these young engineers and bureaucrats, Alphand included, entered into public service and helped France modernize its infrastructure throughout the second half of the nineteenth century. They built roads, bridges, parks, and buildings, and worked their way up into higher levels of French administration. When designing and placing these spaces in the city, studies and diagrams drawn by civil engineering students suggested a similarity between urban planning and landscape design, where plans and diagrams of smaller areas provided a 'view' of buildings and streets.

As they became part of the French bureaucracy, these budding engineers worked for various prefects throughout the July Monarchy and Second Empire who prioritized good administration. Rambuteau, probably the best example of this mindset, favoured piecemeal social engineering over major changes, to avoid unnecessary debt when possible. Of course, Haussmann had no such qualms. Later, between 1871 and 1898, the government returned to the notion that both good governance and fiscal responsibility mattered and would ensure a stable, functioning republic. These examples of competence and a well-run civil service allowed leaders to continue to promote Paris as a modern city and the average Parisian welcomed the stability. For the city's poor, however, life revolved around more simple concerns, like feeding a family and going to and from work. The latter became a greater concern for as industrialization proceeded, workplaces, such as factories, were pushed out from the central districts and often located further from home.

Cartography as Control

Income is a significant factor in how one interacts with a city. For the wealthy, Paris was a featured destination and, with Haussmann's boulevards, easier to navigate. Before widespread access to official cartography, the task of guiding bourgeois residents and visitors through the crowded city fell primarily to the private sector. For example, novels and newspaper accounts fuelled interest in the French capital and described its streets, parks, and monuments in detail. 'Writers were a potent force in disseminating Paris both inside and outside the city. They were the guide – the all-seeing eye – for the uninitiated, for provincials, and tourists' (Green 33). As the nineteenth century progressed, travel guides, the best known of these private endeavours, had introduced Paris to many visitors and provided a first glance (initially through narrative, but later via drawings and maps) of what a city looked like: where one arrived (usually via train), and how to move about the city's dense urban fabric. For visitors, guidebooks assumed this role in three principle ways: subject, function, and format. Guides could devote an entire volume to a city, which allowed the author to provide considerable detail on lodging, food, and sights. The narrative served, as its names suggests, to 'guide' the reader through the unknown streetscape. Finally, their small size allowed the reader to carry them easily and consult them in the street. With these booklets purchased and in hand, travelling to an unfamiliar urban space such as the crowded capital became less daunting. Their concise text and detachable maps afforded confidence to newcomers and that information and knowledge changed how people moved around the city. Furthermore, Paris's wide modern boulevards also helped people make sense of an area as they often provided both a terminus reference point (usually a monument or notable building) and welcome open space, which provided a potential *flâneur* a limited panorama and allowed a visitor to get a sense of the area.

With the completion of new projects, such as the boulevards before the 1878 World's Fair, events played an essential role in the mission of making the city easier to navigate and the official drawings of these new boulevards served as models for guidebook maps. Although not the same level of urbanism that took hold in Europe around the time of the First World War, the blend of public diagrams and private guidebooks document an unofficial cooperation between the two. The fairs showcased scientific, technical, and rational successes; all of which were engines of progress and demonstrations of France's success. Via these events, the Republic was able to intensify its urban improvement by relying on state agencies and administrative organization of the territory (Chevallier 38). In addition to providing new streets in various sections of the city, the Expositions Universelles also presented a French urban ideal. Their large footprint, on both the left and right banks, allowed Paris to create temporary streets and monuments without the usual hassles, such as topography, property ownership, and supporting infrastructure. Beyond an urban ideal, the World's Fairs were symbolic expressions of liberal democratic progress that enabled visitors to experience the future vicariously (Levin 39). These sites thus allowed the Republic to practise its own version of a cityscape before any attempt or implementation in the real world.

The prestige of hosting the event also pushed Paris to finish incomplete projects more quickly. As mentioned earlier in this chapter, when it was announced that Paris was hosting the 1878 Exposition Universelle, city leaders immediately targeted the Avenue de l'Opéra for completion. Due to its diagonal direction, the boulevard had been earmarked for the second phase of development under the Second Empire. Yet both city and national leaders agreed that the boulevard would have immense stature upon completion. It would serve as a base of an ornate corridor from the Louvre Palace, past Garnier's finally completed Opéra, to the developing business zone around the Gare Saint-Lazare.[8] As the city readied for the opening, Alphand finished the diagonal boulevard in record time, just two years (1874–1876) (Landau 39). Yet its inauguration the next year in September 1877 by President MacMahon, along with the Boulevard Saint-Germain, finalized quickly the year before, marked the end of an era in Paris as they represented the last two of Haussmann's central boulevards. From this point forward, it would be republican Paris that initiated and pushed any Haussmann-sized transformations out of the city centre and began constructing its own boulevards in the outer *arrondissements*. For a regime that championed political equality, expanding the infrastructure improvements that transformed central Paris into the outer neighbourhoods made good sense. The city government could use the updated city as proof of the benefits which the Republic accomplished for the county. Furthermore, modernized urban space provided an excellent opportunity to demonstrate how the new government was willing to invest in less touristic sections of the city.

Few residents there would argue that the outer *arrondissements* needed better quality and more roads. The ever-crowded city had long been

8 For more information on the debate around the completion of the Avenue de l'Opéra, see Sutcliffe, *Autumn* 50–51.

notorious for being difficult to traverse. Historically, leaders championed large-scale monumental projects, such as the Pont Neuf or Napoleon's Rue de Rivoli, to remedy this encumbrance and transform how Parisians used their city, and those projects initially improved how to move people and goods. By the 1860s, Haussmann's prestige boulevards and avenues orientated pedestrians confused by the older narrow street network and travel guides highlighted and steered readers to them. By the Third Republic, the embellished *voies publiques* provided a novel view of the Paris streetscape. Beyond their role of improving traffic throughout the city, Parisians and their visitors experienced infrastructural triumphs as they considered them *ouvrages d'art* in themselves (Térade 99). These enhanced streets represented modern city living and would eventually replace the remaining narrow dirt roads in the outer *arrondissements* experienced by Parisians for generations and from which rebellions erupted. Of course, the Republic could also use them to prove that it could modernize just as well as Haussmann.

As the Third Republic continued its transformation of the capital city, there was a desire to prove its ability to represent all residents through the democratic process. Historically, governments used cartography to document jurisdiction – even if only on paper. From the sixteenth to the nineteenth century, for example, most European colonizing nations widely used the discipline for that purpose (see Chapter 2). As France declared the Third Republic, the modern map proved a useful instrument depicting control over territory. Cartography helped ground abstract notions of possession by making distant colonial claims more visually accessible. When imagining a faraway space, cartography affirms precise details such as borders, towns, streets, etc. The British provide an excellent example of this exercise. In the nineteenth century, they worked to create a map that would enable them to govern South Asia, considered by many a faraway exotic possession. In many respects, cartography helped construct British India as an object of rule and offered proof to the governing class in London that its empire confidently controlled much of the subcontinent. Via such maps, one can see how the practice and reception of instruments of governance may sometimes be of secondary importance: what mattered for Britain was the creation of a governable space, one that was 'rational and ordered, that could be governed in a rational and ordered manner,' a space that was white, but also male (Joyce 37). In possession of its own colonial empire in Africa and Asia, France looked to accomplish the same feat since its own lands over the water remained a mystery to much of its population. Closer to home, France also needed to understand domestic urban spaces, document its ability to rule, and prove to its citizens that it was capable of doing just that. Although Paris had calmed down considerably since the 1871 Commune, the fear of uprisings and the general 'unknown' remained. So, just as the British turned to maps to help them rule over the Indian subcontinent, the French government employed cartographic documents to identify and order its own capital city, one in which the Parisian bourgeois would feel comfortable. Furthermore, the regime looked for an ordered urban space in full agreement with the laws and norms of the Third Republic.

If maps served as a means of surveillance and governance over a large foreign colony, they could do the same for a chaotic metropolitan city. When a government masters the ability to map its own territory, it gains the knowledge necessary to control it. Accessible and easy-to-understand cartographic representation allows the individual reader to see him or herself within a larger physical area, an aspect that makes cartography, along with larger city and national maps, effective. For example, the real or imaginary 'you are here' icon centres a person on a map in a specific location, but that representation also helps the reader relate to an overall space or territory.

The ongoing relationship between an individual and the surrounding environment became increasingly important as the use of maps and plans for the negotiation of cities grew. Recent large-scale map employment happened via tourist guides and street atlases. The nineteenth-century tourist map, generally an insert within a *guide touristique*, presented the city as a place of standardized entities full of rail stations, parks, and administrative buildings from which the reader could identify specific local surroundings and attractions. Throughout the century, the built-up urban environment on those maps became increasingly uniform. So, a mapped city in one guidebook would be familiar enough to a reader that they could read and understand a different one found in a second guide. This trend meant that travellers began to comfortably reference common symbols, such as a cross for a church or green space for parks, norms that eventually became standard in French official cartography.

Destined primarily for consumers, private publishers produced *guides de voyages*, home to many maps. However, European governments also played a significant role in their own cartographic output. Starting in

1791, the United Kingdom began publishing its triangulated Ordnance Survey. The series documented the terrain and democratized both urban and rural British space in that, unlike many previous city maps, it did not depend on subscribers. Though not necessarily subject to exclusions and omissions for the reasons listed above, the Ordnance Survey was less exclusive than almost all previous maps of the kingdom. More urban components were represented to a larger audience than ever before (Joyce 44–45). In France, the army remained the primary state agency to assume the role of surveying and publishing terrain maps of the entire nation. Similar to Britain's Ordnance Survey, there was no paid subscription: the army sold individual *planches* directly to the public. More importantly, the *Carte d'état major* treated each quadrant equally, regardless of content, and did not focus on cities specifically, although the Paris *feuille* did situate the capital perfectly. Therefore, true French state urban cartography occurred only when a city fell within a pre-existing quadrant. Even without an urban-specific mission, it is perhaps ironic that the *Carte d'état major* served as a base for much of France's urban planning cartography in the Third Republic (see Chapter 4). In the end, the government's presence in cartography helped eliminate risk and the commercial bias (perceived or otherwise) found in much of the nation's nineteenth-century cartography.

For Paris, urban planning usually meant a powerful man imposing ideas onto the city. Under Napoleon III, Baron Haussmann had made a name for himself both in the capital and around the world with the implementation of Paris's new wide boulevards. For Haussmann specifically, but also for other leaders such as Alphand, modernity also meant using these wide streets to transform the French capital from a crowded, dirty city into a modern, light-filled one. Throughout the Second Empire, baron and emperor worked to alter the composition of the city centre and eventually turned their attention to the outer districts as well. In 1862 Charles Merreau, Secrétaire général de la préfecture, called for the creation of a committee to address embellishments in the exterior *arrondissements*. The commission discussed how to standardize this territory within the existing cityscape. For example, the city would have to rename certain streets where homonyms existed.[9] This commission

9 Patricia Parkhust Ferguson's 'Reading City Streets' states that the 1860 annexation of the immediate suburbs necessitated some 450 new street names (392).

worked to transform the urban fabric and user experience of the city. Yet, in reality, much of the committee's time was spent on the mundane aspects of regulating streets. That commission's report addressed

> les termes génériques employés pour désigner les voies, en précisant des termes jusqu'alors usités de manière vague ou aléatoire, comme l'allée ou la galerie – cette application devait être réservée à des passages couverts et réservée aux piétons. Il proposa également l'abandon de termes inadaptés comme le chemin – qui devait s'appliquer à des espèces ruraux – ou la route – qui devait designer une grande voie de communication entre deux parties du territoire de l'Empire –, mais aussi de termes désuets. Tel était notamment le cas du chemin de ronde qui bordait le mur d'enceinte pour faciliter la surveillance du service de l'octroi: celui-ci n'existait plus, car il avait été absorbé dans la ligne des nouveaux boulevards. (Chevallier 87)

The homogenization of the infrastructural vocabulary epitomized the bureaucratic nature of the committee's work, but also shows the amount of administrative effort needed to fully integrate the new *arrondissements* into the city and update them.

Cartographers needed to update documents to keep pace as official names of streets and buildings changed with each revolution. With the fall of the absolute monarchy, people felt that 'les noms de rues servent de moyens de propagande.' The new government hoped to change them. But the unruly nature of the time hampered that task. Of about 900 named streets in the capital in 1794, the regime changed 53 or only 6 per cent (Milo 283). A complete appropriation of Ancien Régime street names by the Revolution never materialized since the chaotic regime lacked both sufficient bureaucracy and time. Decades later, Napoleon III's fall in 1870 triggered the largest adjustment to the capital when the new government announced a 'de-imperialization' of the boulevards, streets, and buildings. On 12 September 1870, shortly after the official declaration of the Third Republic, interim mayor Etienne Arago issued an *arrêté* to establish a commission of 20 people whose task involved revising street names in the capital (see Demory). The Third Republic thus appropriated these buildings and streets for its own cause. This act affirmed the power of a name and helped eliminate remembrances of past regimes. Within a year, for example, the Avenue de l'Impératrice, which took Parisians from the Arc de Triomphe to the Bois de Boulogne, became the Avenue du Général Uhrich.[10] Furthermore, the commission renamed the Place du Château-d'eau as Place de la République and the Rue du Dix-Décembre, the date in 1848 when France elected Louis Napoleon as President of the Second Republic became the Rue du Quatre-Septembre, the date of the proclamation of the Third Republic.

The authorities continued their republicanization of Paris into the latter decades of the nineteenth century. Paris's Service du Plan de Paris renamed and recategorized streets and then documented the updates on maps. This role required the *service* to publish detailed plans of existing streets as well as reporting changes to adjacent lots and showing how these revisions affected older, already established boulevards. The task, technical in nature, also had a second benefit. The cartographic document recorded street-level modifications in all corners of Paris in real time. The Service du Plan's charts and illustrations, loaded with real-world knowledge, helped the government learn about and administer all areas of the city. Furthermore, city agencies provided additional information to contribute to the urban conversation. Plans that accompanied building permit applications, for example, served as fiscal instruments and as barometers of the financial health of neighbourhoods. The goal was not to know the interior of each individually owned plot, but to understand how each building interacted with the city. In addition to amassing data, state workers also concentrated on documenting public space, those areas within the city used, experienced, and enjoyed by the growing bourgeois sector. The city published or acquired numerous *plans cadastraux et parcelliers*, *plans de voirie*, and *dossiers de construction*. One of the most common of these in France were the administrative cadastre maps that demarcated street improvements, where a red line with a yellow edge often marked the border between public and private property (Lemoine 23).[11] Administratively, such maps helped in the adjudication of property

10 In 1875, the city changed the name of Avenue du Général-Uhrich to Avenue du Bois de Boulogne. In 1929, it became Avenue Foch to honour Ferdinand Foch, Maréchal de France, who died that year.

11 Lemoine suggests that, when streets reached the actual operational phase, maps received writing and comments: the collection of street furniture needed to be placed with extreme precision, following the compositional logic of public space (23).

rights, a key component of any functioning democracy, where the rule of law privileges individual rights. In this capacity, official cartography reproduced the state, reinforced state knowledge, and served a dual role: to map and to inform. Parks, squares, and boulevards welcomed all Parisians. If the city could promote them, residents could better locate and use them. As the city built, remodelled, and enlarged its overall surface area under the Second Empire and Third Republic, maps also removed the unknown and showed where pedestrians could now safely explore. This cartography served as proof of effective governance and provided officials with necessary data. In many respects, the maps also portrayed the large-scale works needed by showing individual property lines and how they affected the neighbourhood or the city as a whole.

The *planches* of Alphand's *Les Travaux de Paris, 1789–1889* provide a citywide portrayal of Paris's current situation. The collection represents another prominent example of the state's ability to use cartography to propagate its message. Christian Jacob suggests that the choice of an atlas to portray the changes to Paris made sense: 'Un dispositif qui permet de concilier le tout et le détail,' the format allows for both a global view through various images (*L'Empire* 97). Furthermore, as a series of maps, the atlas provides a symbolic mastering of space, 'un point de vue gullivérien sur les pays, les régions, les continents [...] Son apparition témoigne de la demande nouvelle d'un public cultivé pour ce genre de volume, non utilitaire, mais nécessaire pour compléter une éducation géographique' (*L'Empire* 98). Via this layout, France eagerly demonstrates the breadth of the Parisian urban reform (streets, water, sewers) it completed in the 100 years between 1789 and 1889 (Pinon, *Les Plans de Paris* 112). No stranger to city planning,[12] Alphand's project did not actually serve a specific administrative role (such as earlier examples, like adjudicating property lines, establishing planning zones, etc.), but instead showed progress as it demonstrated what the Republic was able to accomplish in the city.

The atlas affirms the continued development of nineteenth-century Paris. The maps serve what Joyce refers to as a 'mode of

12 Sutcliff describes Alphand's position in Paris after Haussmann. One of Haussmann's lieutenants, Alphand became director of Travaux de Paris in 1878. Regardless of the Third Republic's desire to eliminate the possibility of another Haussmann, he was able to create an authority almost as broad as that of the former *préfet* by taking control of several departments (44).

governance, namely the reproduction and validation of power (in which consciousness might not come into play at all), and the capacity of the state to convince its subjects that in the more intimate parts of their being they were known' (47–48). Similar to cadastre plans, maps of street improvements represented an investment in the city and Alphand provides a striking visual translation of this belief. Via the various plates, his atlas demonstrates that knowledge and power can translate into a positive notion that benefits its citizens and residents.

The *Les Travaux de Paris, 1789–1889* atlas displays the physical growth of Paris over the past 100 years. It also shows, in a coherent manner, the various stages of new roads, building construction, water works, and street repaving. *Planche* X (ten), 'Reconstruction du plan de la Commission des artistes en 1793' shows the famous 'Plan indiquant les rues projetées par la Commission des Artistes,' one of the first large-scale urban renewal plans for First Republic Paris, a project conceived not long after the fall of the Ancien Régime. This starting point for Alphand's atlas acknowledges the long-term relationship between national governments and the capital city. Even though Louis XIV officially decamped Paris for Versailles in 1682, the Revolution brought Louis XVI and his family back to Paris's Tuileries Palace and reaffirmed the capital's political role. The return of the state, its bureaucracy, and accompanying prestige to the city would do more for Parisians than maintaining a symbolic gilded palace 17 kilometres to the south-east. With power returned within its borders, Paris blossomed. Through the multiple maps, Picon saw Alphand's interpretation of the city as 'an organic tissue permeated by pipes and sewers like a giant brain irrigated by arteries and veins' ('Nineteenth-Century' 144). Picon's visual analogy was not fortuitous, for one of the ambitions of nineteenth-century urban networking was to conciliate two central objectives: modernize the city according to the principles of mechanical efficiency and retain its organic and picturesque character, a picturesque quality especially pronounced in the central, medieval districts. This latter attribute became even more of a concern after Haussmann successfully ploughed through those neighbourhoods in the city centre. Furthermore, Picon points out the government's desire for continuity throughout the various transformations of the French capital. In many cases, he felt that continuance was mere fiction. Alphand's reconstitution of the planning process of the Revolutionary Commission des Artistes plan, for instance, was basically inaccurate.

According to Picon, the main reason for this imprecision was Alphand's insistence on the inspiration between the commission's plan and Haussmann's transformations ('Nineteenth-Century' 144). The desire to associate 1889 Paris with both Haussmann, his former boss, and the 1789 Revolution that toppled Louis XVI may seem contradictory, but by doing so, Alphand benefited from linking himself to both the ideals of the Revolution and the positive results of his former boss and mentor without the negative connotations.

To document all of the work, Alphand provides six reference map dates: 1789 (the Revolution), 1854 and 1855 (Haussmann assumed full control of the city), 1871 (the end of the Second Empire), and 1878 and 1889 (Expositions Universelles hosted by Paris). All the dates correspond to maps entitled 'Eaux et Égouts' and 'Voie publique.' To celebrate the current city, Alphand added three additional 1889 *planches*: a 'Mode de revêtement des chaussées dans Paris en 1889,' a 'Transports en commun,' to document 'Itinéraires dans Paris en 1889,' and an 'Édifices publics, Établissements universitaires et scolaires' *planche* which showed public and education structures constructed in the city between 1871 and 1889. Via these plans, he reaffirms the importance of roads, transportation, and buildings. The city's infrastructure portrayed in these pages was to a large extent the work of the Second Empire (Picon, 'Nineteenth-Century' 144). Even though the selection of maps portrays the capital city throughout the nineteenth century, the large number of post-1871 maps affirms that Third Republic Paris is the primary focus of the atlas.

'Les opérations de voirie exécutées de 1871 à 1889'

Alphand's 'Opérations de voirie' *planche* is easy to read and understand even for those who may not be familiar with Paris or maps in general. Cartography fit in with an overall enhanced access to print in late nineteenth-century France. Joyce suggests that the 'rationality of the modern map depended upon its universality' (36). Theorists of the pedagogy of geography emphasized the 'complementarity of atlas maps and wall maps, and the precise nature of the required gaze and types of exercises for which they were suited' (Jacob, *Sovereign* 94). Where can we find Alphand's message of knowledge and control? *Les Travaux de Paris*'s power comes in part from educational reforms started earlier in the decade, but the atlas's very existence indicates that Alphand published it knowing that an audience existed. A small-but-growing population could read and understand cartography and represented a large potential market, one receptive to the mapped messages which the government produced.

Improved access and understanding helped cultivate a cartographic identity for French citizens in two ways. First, maps fostered individuality. They gave freedom and power to anyone who wished to explore an unfamiliar place. Second, the increasingly universal language of maps helped individuals better envision their place within France and French society. Being a Third Republic citizen in 1889 meant being a participant in its collective history and language, both of which differentiated France from its neighbours. National maps allowed individuals to see how their community or department related to the rest of the country (Joyce 44). As a citizen, one had a responsibility to participate in the functioning of the Republic through school attendance and, if male, military service and voting. These interactions introduced the French to each other on a large scale. Once they read maps, individuals could visualize and better understand how they related to each other on a larger scale. When in front of Alphand's atlas, the reader could better identify with the city. Furthermore, the book's subject affirms its importance to the Republic and the nation. Yet the individual map-reading element is equally strong. The named and dated yellow-and-red boulevards allow a local reader to search out and identify his or her own neighbourhood or address, or one that they knew. For city leaders, the completed infrastructure charted a path to a better understanding of the urban space. Therefore, with this map, Alphand begins to remove much of the unknown experienced by both residents and government officials.

In addition to introducing streets both familiar and new to readers, the atlas also explains the considerable amount of time and energy needed to get Paris to its current 1889 state. The 19 years of construction between 1870 and 1889, added to Haussmann's tenure as *préfet* (1853–1869), lengthen this period of participation considerably. Most readers knew of, and most likely experienced first-hand, the vast amount of disruption caused by the Second Empire transformations. However, with much work in the city's centre now complete, its bourgeois resident could finally begin to enjoy the finished product. Yet with a second wave of construction underway, it appears that Alphand remained undeterred by potential inconvenience to the city's residents. Instead of focusing on the negative aspects, his atlas portrayed the works as a brightly coloured affirmation of the Republic's achievements, just what the government wanted. To obtain this message, Alphand employed the latest cartographic technology:

> Les courbes de niveau sont définitivement adoptées après plusieurs années d'hésitation; pour faciliter la lecture du relief, elles restent accompagnées d'un estompage exprimant un éclairage théorique venant du nord-ouest. La carte s'appuie également sur un nouveau réseau géodésique, établi à la fin du XIXe siècle (la Nouvelle Triangulation de France, NTF, 1873–1891). Enfin, elle s'accompagne d'une légende pour faciliter sa consultation par de nouveaux publics. Le découpage géographique et la numération sont toujours en vigueur. Bien que la lisibilité soit améliorée par l'impression lithographique en 10 ou 12 couleurs, la rédaction reste complexe et les mises à jour le sont davantage. Certains signes, trop précis, sont peu clairs à cette échelle. (Pinon, *Les Plans de Paris* 116)

The breadth of colour and geographic découpage helped Alphand fine-tune his message. 'Opérations de voirie' promoted the substantial work and attention devoted to Paris, the same city that had rebelled on too many occasions. The map helped prove that these improvements worked.

When examining the 'Opérations de voirie' map, first impressions suggest that the outer, newly acquired *arrondissements* received most of Alphand's attention after 1871, and closer inspection confirms this point. There is more yellow in the outer *arrondissements* than in the city's centre. Just as earlier maps in the atlas highlight many of Haussmann's achievements, including the *planches* from 1854 and 1871, some pre-1870 projects, such as the Avenue de l'Opéra and the Boulevard Saint-Germain, remain. With their presence, Alphand subtly reminds his readers that the unpopular Second Empire had been unable to finish its transformation of the city centre. The Third Republic, on the other hand, was both willing and able to complete that task successfully. If the rule for maps is 'the more powerful, the more prominent' (Harley 158), the conspicuousness of colour on this *planche* demonstrates not only the amount of time and energy devoted to the outer ring of the city, but also the capabilities of the government in its ability to improve the entirety of the capital city.

To draw attention to newly constructed boulevards and avenues, Alphand employs bright colours. His choice of yellow and red on a muted background emphasizes the expanding infrastructure, especially since neither tint is found naturally in an urban environment. Nineteenth-century cartographers paralleled a general overall shift in printing style, one that recognized the need for bold printed colours, which were simply better able to stand out on their own (Ehrensvärd 143). Brighter colours simply add life to the document and make the city's construction projects the true focal point of the map. So important are the roads on this 'Opérations de voirie' map that they even overshadow government buildings and moments. Both the Eiffel Tower, a recently completed testament to French greatness, or even the stately Arc de Triomphe at the end of the Champs-Élysées, an established monument completed under a previous regime, are hard to decipher.

In its portrayal of the city, both atlas and map attempt to put the constant bourgeois concern of troublesome residents to rest and Alphand presents a city that no longer needed to fear the working-class 'unknown,' the masses that toppled governments in 1789, 1830, 1848, and 1871. The Third Republic worked hard to 'tame' Paris.[13] Yet after almost 20 years of Second Empire calm, the 1871 Commune de Paris demonstrated that even Haussmann's large boulevards could not stop the *le peuple* from rising up and blockading streets. If wider, supposedly non-barricadable streets failed to prevent a Commune-sized uprising, the government would need to find a way to appease this demographic to maintain order. Instead of trying to suppress them, Alphand shows that they chose instead to interact with, and improve the lives of, residents via significant social reforms, such as free and mandatory education and universal male suffrage. Through a physical amelioration of the outer neighbourhoods of the capital in the subsequent 20 years, the Republic began to repair its damaged reputation with the city's working class, most of whom held a negative view of authority after Haussmann evicted them from central Paris. Wider roads, new schools,

13 With its revolutionary heritage, the Parisian crowd remained a force to be reckoned with. Radical republicans (many of whom hailed from the eastern *arrondissements* of the city) wanted to remind the government of its power. The ultimate goal was to 'create a new representation of the Parisian crowd: it was not to be reduced to isolated individuals, but transformed into an assembly of peaceful citizens' (Lehning 61).

and better access to transportation could significantly enhance the public's opinion of the government.

Political change also helped previously ignored residential areas. After the territorial annexation of 1860, each of the 20 *arrondissements* gained equal representation on the city council. This new redistribution of local government eliminated the monopoly on both projects and influence which central Parisians enjoyed. The need to cater to these other neighbourhoods was clear: residents there voted. Starting in 1871, the outer *arrondissements* returned republican politicians to both parliament and city council.[14] Many public works improvements in outer Paris resulted from these bureaucratic shifts in city administration. Under the Second Empire, the great cost and inconvenience of Haussmann's works had for years been a rallying call for republicans opposed to Napoleon III's regime; Jules Ferry's *Les Comptes fantastiques d'Haussmann* (1868) being one of the most prominent demonstrations of the unpopularity of the *préfet*'s authoritative ways and the obscure methods employed to pay for his works. Yet once in power, republican politicians voted to continue many of these works. Even though few of those lawmakers had the desire to continue Haussmann's overhaul unabated, it was in the interest of each representative to include as many new projects in his district as possible. Not only would this continuation enhance property values and beautify neighbourhoods, it would also employ voters in times of economic stagnation (Sutcliffe, *Autumn* 45).[15] In 1889, the clusters of improvements are found throughout the city, though most notably in the southern twelfth and fourteenth, as well as the northern seventeenth and eighteenth *arrondissements*; proof of the extent to which residents there had benefited from Alphand and the republican regime.

14 Michel Offerlé states that 'le pôle économique des classes supérieures est beaucoup mieux représenté au Conseil municipal de Paris ou à la Chambre des députés.' But between 1889 and 1914, 'les classes moyennes et populaires passent de 19% à 29 % au Conseil municipal de Paris, de 11,5% à 20% pour les députés [à la Chambre des Députés] parisiens' (682).

15 One of the main criticisms of Haussmann's works was that he was using his massive building projects as a type of social programme. The transformation of large sections of the city required an extensive labour force. The working-class sections of Paris were full of unemployed men. By creating jobs, Haussmann and the emperor were able to 'occupy' this restless population and beautify Paris at the same time (Sutcliffe, *Autumn* 45).

Alphand's map discourse targeted bourgeois concerns too. The details provided (street names, dates of completion, etc.) allow an anxious bourgeoisie to look at the 'Opérations de voirie' map and the remaining atlas and see that, thanks to Alphand and the city's Travaux de Paris department, Paris had become more legible and better understood by its leaders. Road construction enhanced knowledge of small streets and alleyways in working-class quadrants, the stereotypical origin of nineteenth-century Parisian uprisings, and could begin to eliminate any unease felt by bourgeois businessmen. Of course, building developers welcomed the continued construction as these new roads and avenues also hinted at the erection of accompanying uniform apartment buildings. Throughout the 1860s, the *immeuble haussmannien* and its standardized façade changed the look of central Paris and gave it its 'modern' feel. In fact, le *boulevard haussmannien* became famous outside of France and other cities soon imitated it.[16] For Paris, new roads had multiple effects. Not only aesthetically pleasing, the wide roads quelled fear as they allowed residents to engage the city, provided order to future transformations, and established authority throughout the city. Making previously unregulated, unknown spaces in the outer *arrondissements* legible, and therefore governable, involved 'untwisting the winding alleys, and smoothing out the irregularity of the space of the courts and the wynds' (Joyce 55). The continuation of the haussmannization-style ordering into the outer districts would further enhance the overall Paris aesthetic.

Beyond documentation, colour reaffirms the map's subject and 'Opérations de voirie' reveals the culmination of city enhancement under the republican-led Directeur des Travaux. For all the attention paid to the outer districts, Alphand still needed to complete a few high-profile unfinished projects in central Paris. On the Right Bank, Alphand documents the completion in 1877 of the Avenue de l'Opéra, discussed earlier in the chapter, and remodels central right bank streets, such as the Rue Etienne Marcel (1885) and Rue du Louvre near Les Halles (1888). One of the most significant endeavours on the Left Bank is the Boulevard Saint-Germain. This road, which forms a *de facto* east–west *croisée* in the southern half of the city and serves as an alternative to the Rue de Rivoli, had been left incomplete after

16 Algiers and Barcelona are two such examples.

Haussmann's downfall. The Republic recognized its role in moving people and goods through the left bank and even used the remainder of the loan used to pay for the Avenue de l'Opéra to finish it (Sutcliff, *Autumn* 51). Two years later, in its desire to connect the Boulevard Saint-Germain to the Place de la Bastille on the Right Bank, the government completed the Boulevard Henri IV (1879), delayed too for its diagonal trajectory, which also included the only non-parallel bridge (opened in 1876) over the Seine.[17]

Republican Improvements to the Outer Arrondissements

On a map promoting road construction in the Third Republic, which section best attracts the reader's gaze? Although the government spent resources to fine-tune and complete the centre, the 'Opérations de voirie' map shows that it is the outer districts that garnered most of the government's attention. The residential quarters, generally overlooked during the Second Empire in favour of monumental, governmental, or commercial zones, now received much more attention. Alphand highlights numerous projects in the fourteenth *arrondissement*. One of the most significant is the 1878 completion of the Avenue Montsouris, the continuation of the long Boulevard Raspail. Similar to the Avenue de l'Opéra on the Right Bank, this pivotal boulevard served as a vital link between central Paris and neighbourhoods and communities to the south and south-east. The long, yellow, easily noticeable thoroughfare terminates at the Parc Montsouris, a welcome oasis for the neighbourhood's population. The green park next door is offset by the surrounding yellow-and-red roads and pink infrastructure, all of which stands out on the tan-and-grey map. The tint evokes the grass and trees there and draws area residents, as there are few other parks nearby.[18] Even though yellow roadways have cut through the Parc Montsouris, the green patch of land was an improvement over even more pink for the residents on the edge of the built-up city.

17 Sutcliffe estimates the cost of building the Boulevard Henri IV at 8 million francs (54).

18 The nearest green spaces to the Parc Montsouris are the Cimetière de Montparnasse, the gardens of the Clinique des Aliénés, and the grounds of the Observatoire.

The Republic completed other *percées*, long boulevards that cut through the dense city, built using the same method employed by Haussmann. The city obtained the land, demolished existing structures, and built anew. A great number of Haussmann's *percées* were actually finished by his successors (Lemoine and Mimran 27). In 1875, the government finally finished the Avenue de la République in the city's eleventh *arrondissement*. Other significant ventures that same year were the Rue Ordener to the north of Montmartre and the Rue d'Alésia, a long east–west boulevard that crossed the fourteenth *arrondissement* near the Parc Montsouris. In 1886, the Boulevard Raspail and the Rue Beaubourg were completed in the centre of the city. Via these avenues, the Republic was able to distance itself politically from the previous regime but also, since they appear on the map, to appropriate many of the boulevards and other endeavours as its own.

The Republic initiated projects as well. Counterclockwise on the map, starting in the south-west, the Buttes-aux-Cailles in the thirteenth *arrondissement* near the Parc Montsouris is another section of Paris that received much of Alphand's attention. Traditionally this hilly, village-like area of Paris lacked accessibility to the rest of the city. Yet by 1889 many of its small pedestrian streets had been upgraded to paved ones. The twelfth *arrondissement*, another area of growth in terms of population and industrial capacity, is perhaps the second heaviest coloured area of the map and Alphand takes pains to highlight the amount of work performed there. Just south of the Gare de Lyon, near the various rail lines, is a large *entrepôt* (warehouse) district initially enlarged and modernized by Viollet-le-duc under the Second Empire. Compared to the area around the Parc Montsouris in the fourteenth district to the west, the updating here centres not just on improving transportation infrastructure and access to parks, but also on the promotion of Paris's growing industrial capacity and its yellow mass of newly constructed roads.[19] During the Second Empire, Napoleon III provided the requisite stability for the industrial revolution to continue in France and for the nation to prosper; the Third Republic needed to encourage economic growth as well. At the start of the Republic the *entrepôt*, next to the Seine where it enters the city limits, occupied

19 The construction of the Entrepôt de Bercy had actually commenced under Louis XVIII. Located outside the city limits, the large *halle aux vins* became one of the largest wine wholesalers in Europe (Fierro 453).

a relatively thinly populated area of the city. Its rehabilitation would help Paris's factories and businesses import and export products more easily. A transition from a sector that warehoused goods into an industrial zone, one that would create more jobs and expand the tax base, was supported by the city. The large mass of yellow here confirms the Republic's commitment to business and industry and by 1889, Alphand's Travaux de Paris office had extensively modernized the area. We can see Alphand's influence elsewhere on the eastern, working-class half of the city too. He improved neighbourhoods in the areas north of Père Lachaise cemetery. The Avenue de la République (1889), the Rue Sorbier (1879), and the Rue Belgrand (1876) link the Place de la République to the newer infrastructure on the edge, near Thiers's wall, an area that still lacked significant government involvement and remained less developed than other neighbourhoods.

To the north and west, both the sixteenth and eighteenth *arrondissement* vie for the next largest amount of yellow-marked improvements on the 'Opérations de voirie' map. Development in the north-central eighteenth *arrondissement* has significance on Alphand's map due to its topographical composition and traditional lack of investment from city leaders. The hilly nature of Montmartre, a small village-like neighbourhood similar to the Buttes-aux-Cailles to the southwest in the thirteenth, hindered any major plans for renovation. Prior to the Third Republic, the area had not yet been a primary focus of the city's attention, although streets and boulevards surrounding it (Boulevard de Clichy to the south, Boulevard Barbès to the east, and Rue Ordender, which would appear to the north) were proposed or finished under Haussmann. Until 1875, when the government laid the first stone of the Sacré-Cœur, there seemed little appetite to expand Haussmann's boulevards in this neighbourhood.[20] Before the basilica's construction had begun, the area had hosted no major organized event or received any significant landmark on the scale of the rest of the capital. Yet as a newly created hub of bohemian life, centred on the 1889 opening of the Moulin Rouge along with other cabarets near the Place Pigalle, more and more Parisians and visitors were venturing

20 David Harvey describes how the committee overseeing the construction of the basilica could lay the first stone without causing much controversy. In the end, a much-scaled down ceremony to lay the first stone passed without incident (*Modernity* 334–35).

to the area each year and it was time the city reacted. Development in Montmartre dovetailed the original mission of controlling and knowing the entire city. To accommodate the new traffic, the city placed large wide boulevards on its northern edge. There, the yellow Rue Ordener, Rue Damrémont, and Rue Campionnet (all completed in 1875) form a section of a border and join with Rue Caulaincourt (1882), Rue Custine (1876), and Rue Carpaux (1882) to finish a golden upgrade of the streets on the northern edge of Montmartre. The northern half of the *arrondissement* now contained transportation routes for people, goods, and tramlines, all of which enhanced the ability of its citizens there to experience and explore other sections of Paris and allowed others to visit the district. Continuing west to the sixteenth, it is not surprising that Alphand highlights achievements in the area across from the Champ-de-Mars. This hilly western area of the city, home to the Trocadéro Palace, was an integral component of the Exposition Universelle site located just across the river and had been the site of many projects.[21] On the edge of Paris, the demographic of this western *arrondissement* was disproportionately bourgeois, but Alphand recognized the importance of the river and the city completed two long boulevards (the *quais* de Grenelle and d'Auteil, 1878), on each side of the river.

Alphand's infrastructural improvements employ other colours as well. Eugène Belgrand's Réservoir de la Vanne (1873), in the fourteenth *arrondissement*, was one of five principal reservoirs in the capital that stored and supplied drinking water for the city's growing population. Its blue colour complements the green parks but also reminds readers that Paris led many other European capitals in the amount of available drinking water per capita and fits within the mission for a republic that worked to improve the health and well-being of its citizens. Hygiene had long held a special place in core republican values.[22] The neighbouring Clinique des Aliénés off Rue d'Alésia benefited from the large number of doctors in republican ranks, doctors who, Conklin states, 'were used to thinking of the nation in medical terms' (59). Its

21 Napoleon I's son was to receive a Palais du Roi du Rome before the city built the 1878 Palais Trocodéro.

22 Alice Conklin's *A Mission to Civilize* mentions that it was only natural under the Third Republic to stress hygiene as part of its *mise en valeur* and, by extension, its mission to civilize its colonies (59).

up-to-date medical care bettered the lives of French citizens. The calm green parks and blue water contrast with the energetic yellow-and-red road projects. The large Bois de Boulogne and Bois de Vincennes were jewels in Haussmann's Second Empire Paris but remained important to Alphand, who had served as the *préfet*'s Ingénieur en chef des Promenades et Plantations, so the role of parks in daily life remained significant in 1889. With increased access to green space, as demonstrated by the entrance to Bois de Boulogne at the Porte Maillot (1878) and the undated *Voie industrielle* (outside the city) that linked the Bois de Vincennes with the Cours de Vincennes and accompanying tramway and omnibus line, more Parisians had contact with the developed park beyond the wall. The partial sections of the two parks that Alphand included on the map are a lush green and they were still important to the city's administration. Nevertheless, the fact that the 'Opérations de voirie' is unable to include the entirety of the two green spaces offers a subtle reminder that the atlas was published to celebrate the Republic, not Haussmann or the Second Empire.

Beyond Roads

Employed to depict roads, parks, and water, colour is the most noticeable discourse tool of the 'Opérations de voirie.' The background colours seem to enhance the roadworks and parks rather than to stand on their own as important markers or tools of republican discourse. Alphand employs tan to identify the shape of Paris while red and yellow present infrastructural improvements. How might Alphand's map reader thus interpret the pink buildings on the *planche*? As Alphand relied heavily on colour, a much easier element to showcase for people becoming more comfortable with maps than unfamiliar signs, the 'Opérations de voirie' makes minimal use of symbols. For his atlas, Alphand chose accuracy over stylized representation, where icons replace the realistic shape of a building (i.e. apple for a school, cross for a church). So the shapes he used on his map are true and to scale. There are two likely reasons for this decision. First, his atlas included other historic maps of the capital between 1789 and 1889. Since he dedicates 'Edifices publics, établissements universitaires et scolaires construits de 1871 à 1889,' *planche* 16 of 16, to buildings constructed in the 1871–1889 time frame, any such structures on the 'Opérations de voirie' serve more as landmarks for the completed roadworks than as objects to be identified and promoted. The 'Edifices publics' map concentrated on buildings constructed in the early years of the Republic, thus eliminating the need to differentiate them on this particular *feuille*.

Yet buildings are both visible and necessary in 'Opérations de voirie.' For this document, the identifiable structures serve primarily as anchors or reference points for the many newly constructed boulevards. New streets did not exist in a vacuum and nearby buildings could help identify where a recently constructed avenue terminated or explain why the city built numerous streets in one location. Normally, when walking through the city, the height of buildings and monuments captures a *flâneur*'s initial gaze. A new boulevard or avenue opened up the city more. It also facilitated the traversal of crowded sectors, the ones in which many of the city's largest buildings were located. The primary interest for the average Parisian was easier movement and interaction with the city. For that reason, yellow, an abundant but bright colour on the *planche*, overshadows Paris's pink buildings, which blend in with the general built-up city. The outline for the Gare de Lyon, for example, situates new roadworks near Ivry in the south-west corner of the city, instead of championing ease of transport in and out of the capital. The normally busy Gare du Nord and the Gare de l'Est similarly position the Rue de l'Aqueduc (1877) and Rue Demarquay (1885), but the stations appear calm and quiet next to the new roads, not boisterous or dirty as they actually were. Large recognizable government or private structures therefore help locate new road construction in the centre of the city. Alphand updated the outline of the Louvre after the 1871 Commune destroyed the Tuileries Palace. Of course, the destruction meant that the city could build the Rue des Tuileries where the former royal residence once stood, so both the western edge of the Louvre and the Tuileries Gardens have yellow roadworks to enhance access between the Avenue de l'Opéra and the Seine but also to remind readers of the damage left by the Commune. The current government's ability to repair the Commune's destruction was a point of pride for Alphand.

Accurate and comprehensive cartography represented one aspect of a confident republic. A model capital city that epitomized republican values was another. Haussmann's Paris was a source, product, and symbol of imperial policies for managing change and there were certain continuities following the Second Empire. First among them was the transformation of the city from an imperial into a republican capital. Upon the annexation of 1860, a large amount of territory and new residents joined the French capital and following the Franco-Prussian war (1870–1871) and the Commune, the city began to shift its interest towards the newer *arrondissements.* To prove its commitment, the government erected various monuments to republican rule throughout the nation, which also appear on the map. The *mairie* was an example of this phenomenon. The town hall, one of the most prominent community-level illustrations of republican infrastructure in the French state, served as the initial point of contact for the average citizen when interacting with the Republic. As part of an overall constitution of city government, the construction of these administrative structures in the outer sectors proved the state's commitment to balancing the attention already paid to the central neighbourhoods. The change from 12 to 20 *arrondissements* in the city required a miniature building spree of town halls. The oldest of the post-annexation town halls, finished in 1860, could be found in the first *arrondissement.* The remaining *mairies* were built either during the Second Empire (in the third, fourth, seventh, eighth, eleventh, and fourteenth *arrondissements*) or started and/or completed (if started under the previous regime, such as those in the twelfth, thirteenth, fifteenth, sixteenth, eighteenth, nineteenth, and twentieth) during the Third Republic under Alphand's leadership of the Travaux de Paris. The government needed these edifices both to represent the Republic and to serve the neighbourhoods and local residents. To enhance these buildings' stature and demonstrate their importance to the republican mission, the government commissioned architects, such as Nicholas Bailly, to design imposing structures to project importance.[23] Ironically, those town halls completed in the Third Republic often required new roads as well, which improved access to them and their government services. However, the brightness of those newly built streets overshadow the new public buildings on 'Opérations de voirie.'

As the government added administrative developments to the cityscape, political obstacles, such as General Boulanger's populist rise (1888–1889) reminded residents and leaders that their newly found political stability could be disrupted despite all these physical

23 Besides the *mairie* of the fourth *arrondissement,* Bailly built the Tribunal de Commerce and the *lycée* Saint-Louis.

improvements. The government therefore needed to enhance its security profile in the city. One effective way to demonstrate its commitment was to create additional barracks. Beside the three Parisian ones already under construction at the start of the Republic, by 1884 there were 18 more throughout the city. These garrisons held both members of the *garde républicaine* and firemen in a joint operation started thanks to a 5 December 1866 decree which brought these two branches of enforcement together under the authority of the Ministre de la Guerre but maintained by the city. Due to its central location and grand stature, the Caserne de la Cité (1863) was meant to be the flagship barracks and to serve as testament to the government's prioritizing of maintaining order in the city. But following a fire in 1871, the government repurposed the building and it became Paris's Préfecture de Police (Chevallier 77). Although there was little need of large-scale troop deployments in Paris after the Commune, the mere presence of these buildings further solidified the commitment made by the government to maintain law and order and to encourage growth. Alphand portrays the *casernes* that would keep the city safe in the same manner. Two of the most prominent and centrally located of these are the Caserne on the Île de la Cité and the Caserne Napoléon, adjacent to the Hôtel de Ville. These were important, but known and both can be found near yellow roadwork improvements, which helps highlight their positions in the city and ease of access if the personnel are needed to help maintain or restore order.

With government structures and institutions in place, the built-up city's expanse of pink shows the extent to which Paris continued to grow as it now abuts Thiers's wall. By 1889, the city is full. Instead of identifying available vacant space, the central white areas on the *planche* highlight memorable buildings such as the Louvre, the Palais Royal, and the multiple rail stations. This use of non-colour allows these structures to stand out more in the crowded capital than medical clinics or district city halls. Further away from the city centre, white space helps distinguish some of the city's other important structures from the background and filled city. The abattoirs and the *Marché aux bestiaux* located in the northern La Villette neighbourhood are two complexes of tremendous value to nineteenth-century Paris.[24] Built

24 For an in-depth look at how nineteenth-century France treated hygiene and modernity, see Vanessa Schwartz's *Spectacular Realities*.

in 1867 to consolidate multiple slaughterhouses scattered throughout the city, they played an important role for the capital. With emptiness around them, these two compounds appear more prominent on the map and Alphand uses them to note the French capital's ability to sustain itself as both structures affirm Paris's ability to process and distribute food. Modern and efficient, the sizeable slaughterhouse and stockyards together covered a significant amount of territory on the city's edge. Perhaps most important for these two icons of food distribution, Parisians could 'see' where their meat was processed when looking at his map. Les Halles, immortalized in Zola's *Le Ventre de Paris* (1873), was located in the city centre, where map readers would likely have, at some point, interacted with the centuries-old, but recently rehabbed, marketplace. Its white surroundings help it stand out in the crowded core. Modernity necessitated that large cities contain and rely on operations that could feed their populations. For Alphand, the mere presence of both stockyards and slaughterhouse, even if pushed out to the edge of Paris, represented up-to-date infrastructure and served as a point of pride. The well-utilized rail lines suggest the use of freight trains, which would negate the need for wide boulevards. So, even though the city relied on the two buildings, few bourgeois residents would be required to venture out there in person.

As French industry continued its ascent in economic importance, the government continued to champion it, as demonstrated by the outline of the Palais de l'Industrie just off the Champs-Élysées, in the city centre. This large pavilion hall, built for the 1867 World's Fair, served as a monument to French industry and housed various exhibits, many of which promoted French industrial might. This modern marvel would also remind readers of Paris's ability to host successful large-scale events. Expositions Universelles in 1867, 1878, and 1889 placed the French capital on the world stage. The Avenue d'Antin, lengthened in 1875 from the Champs-Élysées to the Église Saint-Philippe-du-Roule, would help visitors navigate the vicinity.

Modernity as a concept helped elevate Paris to the forefront of the continent in terms of business and infrastructure. Yet modernity could be a problem in its own right, as Parisians often reacted negatively to it at first. Residents criticized the Eiffel Tower, one prime example, since it did not 'fit in' with Paris's 'modern' aesthetic, and it is virtually non-existent on Alphand's map. The monument's small footprint allows Alphand to hide the structure among all the Exposition's pathway development instead. Since the main structural work ended in March of 1889, a sizeable tower would have been present as the Service des Travaux compiled its data for this map. However, Alphand chose not to include any of the four base pillars here. By doing so, he severely downplayed height for the significant ground-level preparations needed to accommodate fair goers.

Historical buildings contrast with technology and industry but, as integrated components to the Paris cityscape, they too receive some prominence on Alphand's map. The map's white background enhances two historic military structures: the Invalides hospital and the École Militaire. Louis XIV's Invalides Hospital, still in use at that time, cared for French soldiers, many of whom were injured in the Franco-Prussian war. In 1878, to create and educate future leaders, the army requisitioned Louis XV's École Militaire, partially abandoned after the Revolution and neglected during the Restoration, to house the École supérieure de guerre. The map affirms the extent to which the Third Republic still recognized the importance of a strong military to maintain order in the capital, to grow its burgeoning colonial empire in Africa and Asia, and to show readers that it remembered the recent German siege and was prepared to defend the city, if needed.

Most notable, perhaps, is the portrayal of Paris's border. Alphand's pink shows to what extent the city pushes up against Thiers's fortifications, which are presented in a light-grey tint. Alphand's depiction of the wall suggests a transition from a primarily defensive to a jurisdictional role in the eyes of the government. Up close, the fortifications provide an obvious border between the city and land outside. Their primary function, defend against foreign adversaries, had also mandated a 250-meter zone *non-aedificandi* of adjacent open space. This gap, originally reserved for military purposes, had been neutralized and now served as open space on the city's outskirts. Alphand's green border on the outer side of the grey wall suggests that the city accepts this role for the space. Perhaps most important, the wall documents the extent of Paris. Besides the two *bois* and the unnamed suburb to the city's south-east, there is a significant lack of infrastructure outside the wall. The complete absence of representation elsewhere, perhaps due to constraints of space imposed by the physical size of the map, is remarkable. Alphand severs roads and rail lines almost as soon as they cross over into neighbouring communities. Is this the bourgeois view of the suburbs? If so, on this map they simply do not matter. The city of Paris itself, capital of the French Republic, is the map's subject and monopolizes both the map and the government's attention.

'Opérations de voirie' highlights Paris's wide boulevards with the largest text size. Yet the most prominent, visible font on Alphand's 1889 map is found not in the boulevards, but in the open spaces. The influx into Paris of provincial French seeking work throughout the Second Empire and the Third Republic generated severe crowding in many parts of the city and space was precious. The Bois de Boulogne and the Bois de Vincennes were by far the most significant open expanses and, as such, garner the most prominent font on the map. The Champ-de-Mars and the Esplanade des Invalides, integral parts of Paris's 1889 Exposition Universelle, occupied equally sizeable undeveloped terrains in central Paris and, along with the Seine, the city's unofficial divide and the Exposition's 'main street,' receive a large print font.

Both Haussmann's Second Empire boulevards and the Third Republic yellow embellishments appear with a print font. As a result, Alphand's hue favours his street additions and leaves the pre-1871 boulevards white. Following normal cartographic convention, the presence of colour suggests importance as it draws the reader's eye. It is perhaps ironic that this plain background presentation renders the black type on Haussmann's streets easier to read than those with both font and colour. One can decipher the older, better known boulevards much more easily than the newly constructed ones. Furthermore, as the size of the street diminishes, so too does the font. From afar, the names of the smaller, less important streets become more difficult to read. The font-size hierarchy of Alphand's 'Opérations de voirie' further focuses the reader's attention on the broad boulevards that made the city famous. For a city hosting an international event, these avenues would be as much a draw for domestic and international tourists as the fair itself.

By 1889, many of Haussmann's wide boulevards housed omnibus and tramway lines. This monopoly on municipal transportation, formed under the Second Empire and continued into the Third Republic, transformed the way people moved about, used, and knew the city.[25] With proximity to places of employment less a factor than in

25 In an effort to better organize the service, Haussmann combined the ten different omnibus companies into one large Entreprise générale des omnibus in 1855. This became the Compagnie générale des omnibus, which possessed a monopoly on public transportation in the city (Fierro 1032).

the past, citizens of the outer *arrondissements* now had more flexibility in choosing jobs and housing. Public transportation could now carry them to all corners of the city as the omnibus company followed both the city's population and new streets as it expanded service outwards. By 1873, the city introduced rail-based tramways, which could carry more people and slowly replaced the horse-drawn omnibuses. Even larger wagons, designed to hold between 30 and 40 passengers, appeared between 1880 and 1889. These new apparatuses effectively ended omnibus operations (Fierro 1032). Perhaps indicative of their declining popularity and use, Alphand chose not to include the omnibus lines in his map of 1889. The Inspector General did, however, include the ever-popular tramway lines. Drawn in with red ink, they are easy to spot on his white boulevards. Yet, either deliberately or accidentally, the red tramway lines cross out many names, as if Alphand had wished to 'correct' his map. These red lines obscure Haussmann's work, denying them their rightful place on the map and providing the Third Republic with another opportunity to appropriate the wide boulevards for its own use. Alphand's map suggests that, while Haussmann and the Second Empire radically transformed the medieval city centre into a bourgeois playground, the Third Republic did the same to make the bourgeois city into its republican capital.

Conclusion

Paris often represented a challenge to regimes in power. The turbulent nineteenth century upended the city and government too many times. In its desire to move past revolutions, the Restoration, and two Napoleonic imperial regimes, the Third Republic needed its capital city to be an example of a modern French urban space with wide boulevards and uniform facades. During the Second Empire, Haussmann had mollified and updated the once-dangerous city considerably. However, after Napoleon III's 1870 fall, what the Third Republic needed above all was a tranquil capital, one in which business and industry could thrive and help grow the economy. Although not initially supported by all politicians, by 1889 the republican form of government succeeded and greatly changed French society over the course of the Belle Époque. Education and voting, two prime social reforms helped create active citizens. Yet, economically, all were not equal. As recently as 1871, the Commune reminded leaders that, if not addressed, the city's poor could still rise up and disrupt governing. With a representative form of government now established, residents from all neighbourhoods now had more of a voice in how the city and country would be run.

In its mission to modernize, Alphand's Travaux de Paris department worked to update the outer *arrondissements*. As a result, his *Les Travaux de Paris, 1789–1889* atlas documented not only how much of the city the government now knew, but also how far the city of Paris had grown in the 100 years since the storming of the Bastille and the start of the French Revolution. His brightly coloured 'Opérations de voirie exécutées entre 1871 et 1889' *planche* specifically exemplifies these data points. The roadworks had to extend boulevards into the outer *arrondissements*, which involved surveying and obtaining information ahead of time. During the construction process, workers had to demolish buildings to build the new roads. In doing so, they were able to accurately document and interact with the residents living in those once-neglected neighbourhoods.

Beyond the current state of Paris, Alphand's 'Opérations de voirie' map sheet also celebrates the centennial of the Revolution. His 16 *planches* demonstrate the city's progress since 1789. However, the secondary, more powerful message is not that the Republic is strong and growing, which in the eyes of Alphand was truthful, but that the Republic has finally conquered Paris. Alphand finished the work that Haussmann could not; primarily the prestige boulevards that the city wanted and needed. But the bulk of his work, completed under the republican Travaux de Paris, took place in the newly annexed outer neighbourhoods. Thus, on the map, the ultimate goals of Alphand's office are to fix neglected neighbourhoods and better know the city have been achieved. Stability, either under the authoritarian Napoleon III or the democratically elected Third Republic Chamber of Deputies, helped the nation much more than large military victories. The knowledge gained as demonstrated here shows a confident secure city, one that could foster the creativity that would bring about the Belle Époque.

Adolphe Alphand's status and influence changed Paris considerably, but they also resulted in a stranglehold on the planning office of Paris that hindered new ideas, and the office eventually stopped large-scale intervention into Paris planning. After his death in 1892, the city moved to democratize the urban planning process. Reversing Alphand's consolidation, the city divided up various departments and several highly qualified specialists, such as the architects Joseph-Antoine Bouvard and Louis Bonnier, took over (Sutcliff, *Autumn* 220). Alphand, very much a carry-over from Haussmann and the authoritarian ways of the Second Empire, was able to continue the *préfet*'s ideas but also to appropriate them as his own and work productively with the republicans who had opposed his former boss. Of course, many of those same politicians now benefited from having new construction in their own districts. After almost four decades of Haussmann and Alphand, Paris seemed to lose its appetite for providing strong men with the power to transform and update its urban fabric until after the First World War. The region gained new residents and the capital continued to build. Eleven years after the publication of *Les Travaux de Paris, 1789–1889*, the 1900 World's Fair, another large event, brought to the city the *métropolitain* underground railway, the Pont Alexandre III, and the Grand and Petit Palais exhibition spaces. After a period where it had lain dormant, the capital would see a return to the monumental construction.

The first 20 years of the Republic helped set the tone of the Belle Époque. The Commune was the last great uprising of the nineteenth century and, following Alphand's modernizing efforts, leaders felt more comfortable with their capital. The government now knew the city well and the bourgeoisie no longer had reason to fear residents inside the wall. The 'Opérations de voirie' map champions the city over its suburbs, but with more growth, Thiers's wall soon became not a barrier to foreign enemies but one to expansion and national leaders needed to address this challenge. The city itself, in control of only the territory inside its own border, could not survey and map the land outside that administrative line. Other national agencies, such as the army's Service géographique, which published the many *cartes d'état-major* and a subsequent *Carte de France*, would assume that role and map Greater Paris.

Chapter 4

A New Way of Seeing Paris

The Service Géographique's *Carte de France*

> Des pensées de découragement lui venaient; mais André était persévérant: au lieu de se laisser accabler par les difficultés qui se présentaient, il ne songea qu'à les combattre. Tout à coup il se souvint d'avoir vu dans la chambre du garde forestier une grande carte du département, pendue à la muraille: c'était une de ces belles cartes dessinées par l'état-major de l'armée française, et où se trouvent indiqués jusqu'aux plus petits chemins.
>
> —Fouillée, *Le Tour de la France par deux enfants* (1905)

In the Third Republic history and geography text *Le Tour de la France par deux enfants*,[1] famous child adventurers André and Julien hope to cross into France from their native Lorraine in search of their lost uncle, an adult who could help them obtain their French citizenship after the annexation to Germany. As the two orphans prepare to traverse the new border, their guide is injured and the boys worry about how they will proceed. At the last minute André, the ingenious older brother, turns to a wall-mounted *Carte d'état-major*, an official army map in order to navigate across the border.

It is ironic that the French army, so cartographically deficient and easily defeated in 1870 by the same Prussians the youths wish to flee, now provides their eventual means of escape. The self-proclaimed 'meilleur élève' in German-occupied Phalsbourg eagerly employs his map-reading skills to traverse unfamiliar territory and enter 'la France

1 Maps first appeared in the 1884 edition of Fouillée's *Le Tour de la France par deux enfants*. For a broader discussion of the role of cartography in the textbook, see Chapter 2.

glorieuse.' The geography inside post-1884 *Le Tour de la France par deux enfants* suggests that leaders recognized its cartographic deficit and wished to rectify this shortfall at a time when soul-searching and sober examination of France's recent military defeat called for a change in how they understood and portrayed the nation. The expanding use of maps in the French educational system played a pivotal role in helping the Third Republic get to know its territory better and allowed its citizens to understand the maps they saw in the real world. The ability to identify and view a territory mentally became more widespread in the general population, epitomized here by André and Julien's use of the departmental *Carte d'état major* to find their way into France from German-controlled Lorraine. As a result, cartography offered the government an additional tool to promote the Republic. Although the above scene with André and Julien represents an educational ideal, it foretells a time when all Frenchmen would recognize a map and possess the knowledge necessary to use it in everyday situations.

As maps and their employment became more commonplace in the Third Republic, the sale to the public of the *état-major*, its prominence in the classroom, and use by other government agencies meant that, by the turn of the century, a sheet from the Service géographique de l'armée's map would have been available to – or at least seen by – a growing percentage of the population. In addition, the map's ubiquitous presence and state backing meant that the government could deliver updated views of the country to millions of citizens relatively easily. Perhaps more importantly, Paris controlled the representation of that information. The French army's work throughout the nineteenth century, combined with production and technological advances in map publishing, provided the Third Republic with an effective means to certify its knowledge and authority over the entire nation, to showcase its accomplishments, and to highlight the government's benevolent presence in its citizens' lives.

This chapter will examine the evolution of the army's two geographical mapping agencies – the Dépôt de la Guerre, which created the first *cartes d'état-major*,[2] and its successor the Service géographique de l'armée and its map, the *Carte de France* – as they played an increasingly significant role in French society at the

2 The first sheets from the Dépôt de la Guerre's *Carte d'état-major* appeared in 1833.

end of the nineteenth and beginning of the twentieth century. I focus more specifically the Service géographique's 1906 'Paris' sheet (*feuille* XXIII_14), which presents the greater Seine department and documents the continued changing shape and conditions of the capital city. This particular *feuille*, one of the first nine *Carte de France* map sheets to be published, served as part of the overall introduction to the nation of the modified *état-major*, which the army hoped would be more user friendly and commercially viable.[3] Furthermore, the army's *feuille* XXIII_14 offers a governmental viewpoint of the capital's region well before it would formally create one.[4] On this map, readers witness how suburban industrial growth has pushed the outer edges of the Greater Paris urban area well past Thiers's 1840s wall. No longer portrayed as the same isolated bourgeois 'island' presented by Alphand's 'Les Opérations de voirie' in 1889, the capital's urban footprint now fills much more of the Paris Basin.

This military document marks a significant change from other official Third Republic maps, such as Alphand's, which had all but ignored Paris's environs. He had been able to focus exclusively on the city thanks in part to a general bourgeois bias that favoured Haussmann's transformed city centre over the outer industrial working-class suburbs. Yet, by the turn of the century, urban map reception and consumption had changed. Thanks to educational reforms and a more cartographically proficient general public, the army hoped that the *Carte de France* of 1906 would appeal to more people than just political or bourgeois elites. In addition to a larger potential readership, the army's mission to construct an accurate, easy-to-read final product – one that would correctly portray the current state of Third Republic France – meant that it could no longer maintain its previous blatant favouritism towards the embellished city centre, as exemplified by Alphand. Furthermore, by 1906, the Corps d'ingénieurs-géographes had spent decades surveying the nation. Any obvious exclusions or omissions would be seen as a

3 De Margerie quotes Berthaut, who states that 'jusqu'à présent, les fonds disponibles n'ont permis que l'essai auquel il est procédé pour les neuf feuilles en cours de rédaction des environs de Paris.' In de Margerie's opinion, 'essai' meant 'faute de ressources' ('La Nouvelle carte,' 239).

4 In Chapter 4, I discuss the 1919 loi Cornudet that required a regional planning document for *communes* with a population above 10,000. In Chapter 5, I examine 1932 legislation which mandated an urban plan for the area within 35 kilometres of Paris's Notre-Dame Cathedral.

waste of the military's time and resources, and affect its ability to defend France, especially since a lack of cartographic knowledge had played a role in the nation's 1870 defeat to the Germans.

In addition to its accurate portrayal of the French capital, Paris sheet XXIII_14 exemplifies how the Dépôt de la Guerre and the Service géographique both professionalized and homogenized French state cartography throughout the nineteenth century. This uniformity allowed for a more effective cartographic tool, whereby any citizen who saw the Service géographique map would be able to understand it. In short, once a reader, such as André in *Le Tour de la France par deux enfants*, saw or understood a state-produced map, any subsequent state cartographic exposure would contain enough of the same characteristics that they would have a good chance of recognizing it. Therefore, even simple choices, such as font and colour, had a significant impact on how people viewed and identified with the national terrain.

André and Julien's use of the *Carte d'état-major* to enter France from German-occupied Phalsbourg marked a critical component in demonstrating the benefits of well-rounded instruction and the invaluable opportunities offered to those citizens who could read, understand, and use maps. Thanks to Third Republic educational reforms, the French could successfully read more of them. In the first few decades of the new regime the government had relied on schoolbooks, language, and a common history to legitimate the French Republic before both its citizens and the world after its loss to the Germans. Now with the publication, distribution, and sale of the *Carte d'état-major* and subsequent *Carte de France*, government leaders could include state maps – with their professional, uniform, and comprehensible cartographic portrayal of the nation – to the list of objects that helped them portray and explain France to the French.

La Carte de France

In 1890, President Carnot created the Commission centrale des Travaux géographiques, composed primarily of representatives from various governmental ministries. This new commission's task was to consolidate under one central umbrella organization all the individual cartographic projects taking place in agencies throughout Paris and nationwide at the departmental level. The president believed that this new entity, based in Paris, would better promote the rationale for updating the *Carte d'état-major*, eliminate duplication of jobs, consolidate and improve production methods, and hopefully lower the overall cost of a new nationwide map (Bigourdian 436). The Commission centrale announced that the map, which would be known as the *Carte de France*, 'les feuilles, mesurant 20' centésimales en latitude et 40' en longitude, auront (zone moyenne) 0m,40 sur 0m,55 environ. La carte sera en courbes de niveau et imprimée en huit couleurs; [...] Les types et les dimensions des écritures seront conformes aux principes admis pour l'exécution de la lettre sur la carte d'Etat-Major...' (De Margerie, *La nouvelle carte* 240). In adhering to these standards, the agency continued its homogenized presentation to meet the expectations of the growing French market.

In August 1900, General Berthaut, on behalf of the Service géographique de l'armée, presented initial *Carte de France* sheets to the Congrès national des Sociétés françaises de géographie at its Paris headquarters and the specialized audience welcomed the new map warmly (Vidal, 'La Carte' 114). With initial feedback positive, the army's topographic brigade conducted and published new surveys between 1889 and 1903. As the first map sheets appeared, a few outside groups criticized its presentation of France. The American-based *National Geographic* magazine, for one, felt that the Service géographique's triangulation, heavily utilized earlier in the century, was too elaborate for a map of this calibre. In addition, *National Geographic* felt that the actual territorial depiction, more specifically its artistic production, fell short of expectations. What good is accurate data, it asked, if the final product contained low-grade sketching? (Collet 251). The poor presentation detracted from the *Carte de France*'s usability. The editors did include some positive comments, however. For one, they felt that the Service géographique's surveying had accumulated a considerable amount of accurate and valuable information.

In contrast, the initial drafts of new map project received high praise domestically as they marked a significant improvement over the previous *état-major* documents. The army anticipated that the new *Carte de France* might even help France regain its long-lost cartographic superiority, a worthy goal that still weighed heavily on the Service géographique leadership. Berthaut believed that the upgrade from the old *Carte d'état-major* to the new *Carte de France* was equal to or greater than that from Cassini's map to the Dépôt de

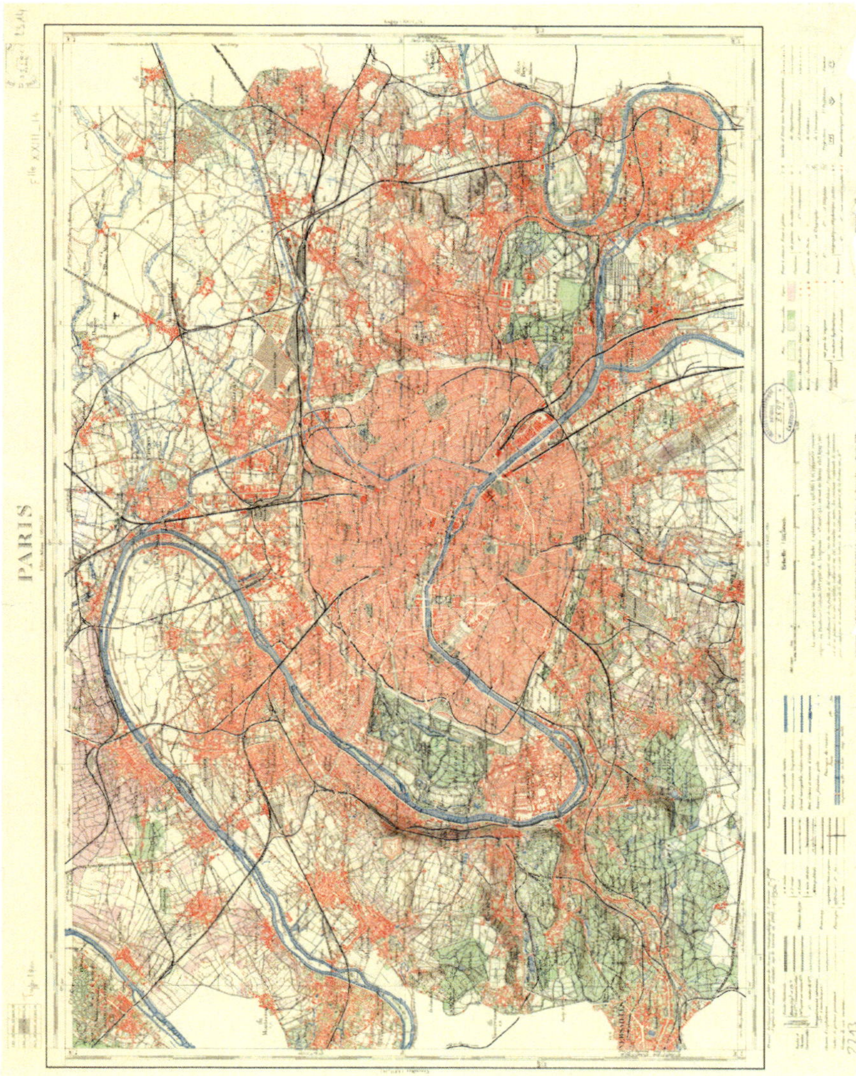

Figure 5. Service géographique de l'Armée, Paris, *feuille* XXIII_14, 1906.

la Guerre's *Carte d'état-major*. He even went as far as to compare the new and improved map of France to Cassini's publication a century previous, which drastically changed how the French viewed and understood the nation.

Other leaders were less sanguine about the new project's openness and accompanying fanfare, especially given recent French military defeats and the fact that upon publication, other nations would gain access to the same valuable information about infrastructure to which the army had access. General Demarçay disagreed. The same concerns did not stop France's neighbours in Europe from publishing their own maps and even if foreign governments could view details of French terrain and urban infrastructure, he saw more benefits than drawbacks for the nation (Berthaut II 266). To him, the mere process of scouting territory, drawing the initial drafts, engraving, and publishing them would provide the army with invaluable data and first-hand exposure of national territory. The newly updated, higher-quality map would also improve knowledge of the current state of France and make it available it to the French population, who could themselves use it for their own commercial or personal needs.[5] With the initial phases of the new map underway, the Ministre de la Guerre decided on 14 February 1903 that the Service géographique de l'armée would produce and engrave the *Carte de France* at a scale of 1:50 000. Following a short instruction period, the corps's topographers were sent out throughout the country for six months. Berthaut called the results of those surveys 'remarkable' (de Margerie, *La nouvelle carte* 238). The government's initial desire in 1818, to present the nation with an accurate cadastre map, had, by the publication of the XXIII_14 sheet in 1906, provided it with considerable detailed information of the national terrain. The government adapted the document to the changing nature of French politics and the economy. Instead of addressing an almost exclusively military clientele, the Service géographique needed the new maps to address private and civil needs to justify its existence and provide revenue to the government.

The army hoped that the nation's private sector would also embrace the *Carte de France*'s economic potential. The presentation of infrastructure, roads, and bridges showed businessmen or factory

5 The *Guide Michelin*, for example, relied on the *état-major*'s cartography for its first maps (Olson, 'Driving a New Tourist' 208).

owners where to find and access natural resources and how to transport them back to factories for development and processing or to markets abroad. In that respect, the map's newer, more accurate depiction of terrain served the public well (de Margerie, *La nouvelle carte* 243). *Communes* could better project and manage growth, a much-needed capacity given the significant nineteenth-century internal migration that resulted in a large influx of domestic 'immigrants' moving to Paris and other large cities.

The Creation of a State Mapping Enterprise

In 1818, approximately 150 years after the Cassinis first triangulated the nation and began producing their pioneering large-scale survey of France, the French army's Dépôt de la Guerre planned a series of survey maps that would be known as the *Carte d'état-major.* The Dépôt's map series project served as a sequel to – but was also meant to improve upon – the work of the Jean-Dominique Cassini (Cassini I) originally invited to France by Louis XIV and his descendants: his son Jacques (Cassini II), grandson César-François (Cassini III), and great-grandson Jean-Dominique (Cassini IV). The Cassinis held a special place in the hearts and minds of the French scientific and cartographic community and the family epitomized French triumph over nature and the unknown. The Cassini map, groundbreaking in its portrayal of pre-revolutionary France, changed how Ancien régime society viewed and understood itself. Following the Revolution, the Cassini maps maintained their position in early nineteenth-century French cartography for two primary reasons. First and foremost, the Cassini family had provided France with its first true national state map, one that was not commissioned by or held in private hands and, in theory – although not necessarily in practice – accessible to all. Their *Carte de France* represented an overall change in the idea of scientific patronage and confirmed the state's role in developing and supporting national projects in the seventeenth century.[6] Second, the Cassini map

6 Unfortunately, when Cassini III presented Louis XV with one of the first sheets of the map in 1756, the king told the cartographer that 'je suis bien fâché: j'ai une mauvaise nouvelle à vous apprendre; mon Controller général ne veut plus que je fasse continuer la carte. Il n'y a plus d'argent pour cela.'

afforded individuals – at least those with access to them – the ability to assemble multiple or even all 174 sheets of the kingdom together at one time in order to create a large detailed map of France, which had been impossible prior to their project.[7] The assembled *feuilles* thus displayed not only the vast size of one of Europe's largest kingdoms, but also enough detail to portray its true shape accurately.

The mere fact that the Cassini family had been able to travel about the kingdom to produce such a remarkable work is noteworthy given the state of Ancien-Régime France.[8] With no major coordinated drive to build infrastructure in the first half of the eighteenth century, the École des ponts et chaussées, founded in 1747, built roads to link Paris with many of France's larger cities. Prior to the 1770s, however, much of the rural countryside remained largely untouched (Duby, *Civilisation* 121). Triangulating relatively accessible French cities, farmland, and coastlines proved fairly straightforward. However, once these easier portions of the national space had been surveyed, Cassini sent his engineers to address difficult, less accessible sectors, such as rural regions and the higher elevations of the Alps and the Pyrénées mountains.[9] In those locations, weather, dangerous terrain, and even simple isolation proved taxing and took its toll on personnel and morale. The Cassinis resorted to paying their engineers various 'primes d'encouragement' in order to keep their workforce from packing up and leaving their posts in these less than ideal conditions, an eventuality which would have caused significant delays to the project (Berthaut 1 52). They overcame many of these issues and by 15 August 1756, the Cassinis produced their first cartographic document. Thirty years later, in 1786, they completed their final surveys with

After receiving this news, Cassini formed the Société de la carte de France, which had 50 members, including the Marquise de Pompidou, and the goal of continuing his work (Pelletier 143–44).

7 Monique Pelletier lists the 175 *feuilles* of the *Carte de France de Cassini*. Sheet 160, for the Île de Noirmoutier, was never published. The island was placed on *feuille* 131 with Nantes (255–65).

8 For further detail and a map on the growth of French roads in the nineteenth century, see Duby (119–20).

9 Berthaut, a critic of the Cassinis' topographical portrayal, lamented the fact that the map's representations of higher elevations were completely insufficient. Citing the example of Pelvoux, a *commune* in the Hautes-Alpes department, he concluded that the Cassini engineers 'ont reconnu leur impuissance et renoncé à décrire le terrain' (57).

the last sheet, number 175-8A 'Audierne – rade de Brest,' published shortly thereafter (Pelletier 265).[10] So the Bourbon monarchy got to see the fruits of the Cassinis' labour in the final years of the Ancien Régime.

After the fall of Louis XVI, French cartography persevered. On 25 November 1793 the Convention transferred the Cassini plates and all 174 map sheets, classified as state property and held by 'le citoyen Capitaine,' from the Société de Cassini at its home in the Observatoire to the Dépôt de la Guerre, home to France's military records, archives, maps, engravings, and topographical services (Berthaut I 59).[11] By formally reassigning the Cassini family's work to the Dépôt, the Convention effectively nationalized state cartography and solidified the military agency's role in any future nationwide cartographic project. However, until any new map could replace it, the Casssini *Carte de France* would remain France's default national map. Upon taking possession of the map plates, the Dépôt de la Guerre's initial inspection uncovered many significant inaccuracies. For example, the army found that some plate-engraved localities appeared up to a league (approx. 3.54 miles) away from their actual position and, in certain cases, the Dépôt failed to verify the exact longitudes of *communes* when employing the Cassini engineers' calculations (Berthaut I 173). In addition to these errors, overall map plate accuracy worsened under the new owners. Every time its engineers discovered a mistake, the Dépôt de la Guerre had to 'correct' the plate manually, usually through additional engraving. Over time, the resulting physical degradation from these adjustments rendered many of the plates useless. Between imprecise measurements, poor engravings, and the subsequent mediocre quality of the modified plates, any long-term usefulness of the Cassini map's fell into doubt, especially since contemporary cartographic portrayals of sectors of France, such as the 1764–1773 *Carte des Chasses*, which mapped the environs of Versailles, were more accurate (Bigourdan 431). Of course, these newer examples had fewer problems since they represented smaller land areas and/or catered to a specialized audience. Regardless of its

10 To view a timeline of map sheet publication under the Cassinis, see Bigourdian (428).

11 The Dépôt also received map plates of conquered territories in Belgium and the Netherlands at the same time (Berthaut I 60).

numerous drawbacks, however, the Cassini map persevered as governmental instability, brought on by la Terreur, French civil war, foreign invasion, and the rise of Napoleon, halted any general discussion of a new nationwide survey.

By the beginning of the nineteenth century, nearly 50 years after the Cassini dynasty published their first sheet, the poor quality of the map and its plates left it largely insufficient for widespread national military use.[12] Napoleon, himself a military general, acknowledged the current state of France's cartography and became the first post-revolutionary leader to propose a new map for France. The emperor's cartographic vision was, in fact, less a defensive concern than a commercial one. As he introduced his 1804 Code civil, a massive overhaul of the French legal system, he realized that the nation needed a tool that would allow his new and growing bureaucratic administration to adjudicate property lines accurately, a difficult task given the current state of French cartography. In 1807, fresh from his victory at Friedland and the Treaty of Tilsit, Napoleon suggested to his Ministre du Trésor that 'un bon cadastre parcellaire sera le complément de mon Code, en ce qui concerne la possession du sol. Il faut que les plans soient assez exacts et développés pour fixer les limites des propriétés et empêcher les procès' (Pelletier 249). The addition of an accurate description of the terrain would complement his regime's substantial administrative and bureaucratic reforms. As an added benefit, such a representation would also provide the emperor with an up-to-date assessment of the entire nation. However, the foreign campaigns in the latter years of Napoleon's empire and subsequent lack of trained military engineers to perform the necessary domestic surveys, his disastrous march on Moscow in 1812, military defeat, and forced exile ended any chance of the emperor producing his sought-after national map during his reign.

Following the 1814–1815 Congress of Vienna and Napoleon's final defeat at Waterloo, the return of the Bourbons to the French throne provided the requisite political and military calm to allow the nation to refocus its attention on updating the national map. In 1816 Commander Maxime-Auguste Denaix, head of Dépôt de la Guerre's newly reconfigured Corps of Ingénieurs-géographes (Corps), became the first post-Napoleonic official to revisit the idea of a new

12 Pelletier mentions that the first of the Cassinis' 181 plates, surveyed between 1749 and 1755, was that of the capital (8).

cartographic survey. The general believed that the nation needed a functional document that could focus on topography and military defence, but which could also address 'tous les services publics; l'ingénieur des mines, l'ingénieur des ponts et chaussées, le géologue, le géographe, l'historien comme le militaire, devaient y trouver tous les renseignements divers nécessaires à leurs travaux' (Blondel 3–4). With revolution and foreign campaigns fresh in Denaix's mind, any future state-supported cartographic project's primary purpose needed to be to serve the army's needs in national defence. If done correctly, however, such an endeavour could also boost the prospects of the kingdom's commercial sector.

Traditional military leaders too spoke of abandoning the Cassini map for a better quality cartographic document. In 1817, Général de division Marc-Jean Demarçay warned of possible drawbacks to France if it did not retire the Cassini map:

> Cette carte serait nuisible, si la guerre arrivait, et vous allez le sentir: si nous faisons la guerre en France, vous concevez que les généraux et les officiers français connaitraient mieux le pays que les étrangers. Vous devez supposer aussi qu'ils recevraient des habitants de meilleurs renseignements que les étrangers. (Berthaut I 264)

Demarçay felt that the state of the Cassini map left France at a disadvantage. First and foremost, its inaccuracy worried him. In any conflict, neither French officers nor their soldiers would have access correct information of the domestic terrain. He strongly believed that the army should resurvey the nation in order to correct the Cassini coordinates and produce new engravings. In truth, 'la carte [de Cassini] resterait toujours insuffisante, à cause de la médiocrité de son fond même, et les travaux dont elle était objet d'avaient pour but que de la prolonger en attendant qu'on pût disposer d'une œuvre nouvelle' (Berthaut I 172). In the 50 years since its publication, not only had the French produced better smaller, localized maps, but the English, Germans, and Russians too now had '[d]es cartes infiniment meilleures' (Berthaut I 266). Given that Cassini's series was France's only large-scale option, the nation needed a new map that could provide its leaders with an up-to-date view of the nation's landscape and infrastructure.

An ordinance issued on 11 June 1817 officially established a royal commission to create a new *carte de France*, one that could be combined with the operations of the general cadastre (Blondel 5).[13] The final product would document the post-Napoleonic borders and provide a cartographic foundation upon which both public and private maps could be based. The next year, in 1818, the Dépôt's corps took the lead in surveying and implementing the royal commission's orders.[14] In a nod to its predecessor's influence, the Dépôt employed the same scale as the Cassinis. The nation, or more specifically the army, had relied on a national map at that scale for decades and its continuation made practical sense. As France moved forward with the project, Louis XVIII also charged a second Commission spéciale du Dépôt de la Guerre to focus on 'des moyens de réaliser [les décisions de la Commission royale] ainsi que de ceux des ressources que l'État pouvait en tirer' (Berthaut I 188). The two groups immediately divided up preliminary tasks. The first team took charge of surveying the territory and the second searched for appropriations to pay for the project. This separation provided helpful flexibility at this early stage and allowed each commission to concentrate on its own task without burdening or slowing down the other.

In order for this new project to be successful and improve the quality of the final product, the Dépôt had to surmount the Cassini personnel shortcomings. To overcome these, it needed a large, mobile, and competent selection of men who could understand geography, employ scientific instruments, and take orders to deploy in less-than-favourable locations. The army felt that the most effective solution was to form its own educated and trained workforce. Under the empire, Napoleon had relied on a quasi-bureaucratic corps of engineers, epitomized by both the École des Mines and the École des ponts et chaussées, to modernize the nation's infrastructure. The Dépôt, hoping to follow the emperor's example, used its own bureaucracy and employed the

13 The commission was made up of 14 members from various governmental agencies and schools, including the École des ponts et chaussées and the École des mines. It also included the Adjoint Director of the Dépôt de la Marine, the President of the Département de l'Intérieur, the Ingénieur en chef of the Cadastre, and three members of the Conseil de l'Administration générale des forêts (Blondel 5–6).

14 Blondel lists the following men as members of the *bureau spécial*: *messieurs* Muriel, officier supérieur d'état-major, Denaix, Jacotin, Bonne, Brousseau, Puissant, Maissiat, Delahaye,. Coraboeuf (all *officiers supérieurs du corps des ingénieurs-géographes*), and Dilhe, secrétaire-général du Dépôt (10).

Corps des Ingénieurs-géographes, a bureau of trained cartographers, to professionalize French cartographic industry.

If successful, the establishment of a sizeable and knowledgeable map-producing labour force would mark a definite change from the relatively small number of engineers the Cassinis had employed a century earlier and help the Dépôt overcome the former's shortcomings.[15] In order to build a true professional band of surveyors, the Dépôt needed a sizeable pool of men from which to select this new army of geographers. It debated how to recruit capable men who could survey in the field, a specialized task at a time when education in France was limited to a small percentage of the population. One option for the Dépôt was to obtain new field assistants from the military itself. However, in order to ensure the consistent quality of its staff, the Dépôt preferred to absorb only high-level personnel and not accept those with the rank of *sous-officier* or below. The Cassini map contained many logistical errors and the Dépôt wanted motivated and detail-oriented individuals to fix them. Its administrators also feared that lower-level officers would require too much supervision, which posed too great a risk to the future map's worth. The army, with a large quantity of men at its disposal, could easily supply the Dépôt with enough recruits, but the Dépôt feared obtaining questionable results if it sent unmotivated or too inexperienced young men wandering around France. Officials cautioned that even one untrained section of workers 'peut nuire à l'homogénéité du travail, à sa qualité, tout en en hâtant la terminaison; elle peut nuire aussi à la réputation du Corps des Ingénieurs géographes' (Berthaut I 228). However, the national government wished to bring in soldiers from lower levels to fill the Dépôt's ranks more quickly. If the agency could accept new recruits at the rank of lieutenant or *sous-lieutenant* aged 20 to 30, the potential new influx would represent a significant opportunity for both it and for the nation. In the government's eyes, 'à cet âge [on] trouve des sujets plus propres à recevoir et à mettre en pratique l'instruction dont les auxiliaires auront besoin; à soutenir les fatigues d'un métier pénible; plus dociles, et enfin d'une émulation plus faciles à stimuler' (Berthaut I 229). Furthermore, if identified and trained early enough, these men would become the next

15 Monique Pelletier describes how Cassini III wished to rectify the insufficient number of trained personnel. He hoped to get up to a total of 34 engineers (156).

generation of *ingénieurs géographes* and prolong the life of the Dépôt de la Guerre's cartographic unit.

The Dépôt acquiesced and accepted all ranks of soldiers into the Corps. The Commission royale considered that 'l'initiation d'un certain nombre d'officiers, soit de troupe, soit plutôt des États-majors, aux travaux topographiques, ne pourrait être qu'utile au bien du service, à l'instruction générale de l'Armée, à la vulgarisation de la connaissance du terrain, etc.' (Berthaut I 228). These recruits bolstered the ranks of the Corps, which hired on average 54 such officers each year. On top of these soldiers, it acquired artists, engravers, and other miscellaneous service employees (Berthaut I 230). The Corp's numbers grew spectacularly. 'On a employé à la carte de France, depuis 1818 jusqu'en 1830, 54 officiers tous les ans; en 1831, et depuis la fusion des officiers d'état-major avec les ingénieurs géographes, le nombre de ces officiers a été augmenté; porté pour 1831 à 83, en 1832 il sera de 90. Des dispositions sont prises pour qu'il soit de 100 en 1833, de 110 en 1834 et de 120 en 1835.' (Bethaut I 264). In addition to engineers, the army introduced over a thousand military personnel (officers, artists, engravers, etc.) into the Corps over the next two decades.[16] This rapid enlargement changed the composition of the agency, but also allowed it to assume the national map project more confidently. Throughout subsequent decades, these new recruits brought with them hands-on experience and provided France with a sizeable and growing cartographic workforce.

As personnel numbers grew, the government reformed the French cartographic agency's administration. In 1831, the July Monarchy moved the Dépôt's Corps into the larger Corps d'État-major.[17] Thirteen years later in 1844, a royal decree reorganized the overall internal composition of the Ministère de la Guerre. That order maintained the cartographic branch of the *état-major*, recognized the importance of the Dépôt de la Guerre, and divided its Direction générale into five sections:

16 Berthaut estimates the total to be about 1,082, which he calculates based on an average of 54 auxiliaries assigned to the Corps des Ingénieurs-géographes for 20 consecutive years (I 229).

17 Berthaut notes that there was a distinction between the two agencies. Originally the *Carte de France* employees were 'temporaires,' and could only switch to a position at the Dépôt de la Guerre after 'de longs essaies et comme avancement' (Berthaut I 272).

Nouvelle carte de France, Travaux topographiques intérieurs, Travaux historiques, Statistique militaire, and Administration et comptabilité. Primarily an administrative move, this separation allowed greater specialization. The legislative action also identified specific duties and advancement and promotion procedures for each section. In addition to the five divisions, the decree also established an official hierarchy of ranks and job posts within the Dépôt and created a level of paid internships, 'des élèves dessinateurs et graveurs admis en apprentissage' for males aged 16–18 (Berthaut I 271–72). This remunerated training provided even more bodies to send out to the field, but also the hope that the young geographers would become the next generation of French cartographers.

The composition and size of this new specialized workforce provided greater flexibility in the Corps's deployment of engineers throughout the nation. Yet even as late as the mid-nineteenth century, the logistics of sending out an army of cartographers into rural France remained a challenge. Outside urban areas, France's vast territory still posed many of the same problems experienced a century earlier by the Cassinis. The French road network, more comprehensive than before the Revolution, nevertheless still failed to provide sufficient access to many remote locales. However, each spring the Dépôt dispatched between 50 and 80 young officer-cartographers throughout the nation to record the territory:

> Ils allaient s'établir chacun dans quelque coin ignoré du territoire, sous quelque cabane informe, et parfois même sous la tente. Là, pendant des mois entiers, actifs comme des missionnaires, ardents comme des apôtres, laborieux comme des bénédictins (les vertus militaires se rapprochent plus qu'on en pense des vertus religieuses), au travail avant le lever du soleil, jamais rentrés avant son coucher, soutenus par un seul sentiment, l'honneur de bien faire et d'accomplir consciencieusement leur mandat, ils présentaient ce remarquable phénomène d'hommes liés à une œuvre ingrate sous beaucoup de rapports, et s'y livrant sans témoins, sans spectateurs, sans l'excitation permanente des chefs et sans l'entrainement de l'exemple des camarades, avec un admirable zèle. (Blondel 28)

These officers epitomized the new État-major personnel. Acting as government emissaries, this group incarnated military values. Their professionalism, drive, and dedication to the project helped make the subsequent surveying and charting of the French territory a success. Blondel signals 20 engineers that the État-major posted near Grenoble, Briançon, Gap, and Die in the Hautes-Alpes department in 1853, not far from a location where Cassini personnel had failed before. In the midst of glaciers and treacherous terrain, this dedicated group spent months stationed at elevations above 2,500 metres to document this section of the Alps. Their zeal and resolve under difficult circumstances resulted in an accurate survey of the difficult topography and represented a triumph of French bureaucracy and science. The Corps's engineers surveyed throughout the Second Empire and into the Third Republic.[18]

With this professional army now covering all parts of the country, Paris continued its reform of state cartographic production. A bill in December 1875 first proposed the formation of a Commission centrale des Travaux géographiques. Two years later, on 20 February 1877, that commission created a body made up of officers, hydrographic engineers, and bureaucrats from diverse ministries under the command of the Ministre de la Guerre, to coordinate the 'geographical sciences.' This new entity would become the Service géographique de l'armée. On 21 March 1890, the Direction du contrôle du Ministère de la Guerre informed the newly formed Service géographique that the budget office would task it to fulfil all ministerial-level cartographic needs (Berthaut II 254–55).

For the state to succeed as a producer of cartography, the average citizen needed to be able to read and understand what it published. In 1875, Frédéric Hennequin, president of the Société de Topographie de Paris and member of the Congrès international des Sciences géographiques, helped broaden the French cartographic industry's impact when he published a textbook primarily to instruct newly formed *ingénieurs géographes*, but also those Frenchmen unfamiliar with maps who wanted to know more about them following the publication and sale of the *Carte état-major.* His *La Topographie mise à la portée de tous* (1875) provided readers with 'les éléments indispensables pour la lecture rapide des cartes et la connaissance des principes de topographie,' all of

18 Bigourdian provides a map of the progress made by the Dépôt de la Guerre in its triangulation and engraving of the *Carte d'état-major* (428).

which could be learned 'en quelques heures' (6).[19] Through his book, Hennequin informs his readers that in order to understand a map and read it correctly, they must know how to combine multiple cartographic components as they pertain to a two-dimensional presentation.[20] Beyond teaching interested parties how to read the uniform state cartography, Hennequin's *Topographie* lists the official colours, taken directly from the Dépôt de la Guerre's guidelines. His readers learn that the *Carte d'état-major* presents *villes*, *villages*, and *maisons* 'en rouge'; waterways (*Eaux*) (which includes *rivières*, etc.) in a 'bleu de Prusse'; and seas (*mers*) in a 'bleu de Prusse mêlé avec une faible partie de gomme gutte' (9). He also matches hues and symbols to diverse categories, such as 'bois, vignes, marais (et marais salants), falaises, dunes et sables.' For man-made divisions on the *état-major*, for example, the Dépôt de la Guerre chose tints to identify jurisdictional boundaries. The limits of international 'états' are in a 'bleu de Prusse foncé,' departmental ones are 'en bleu plus léger,' *arrondissements* are in 'rouge,' cantons in 'vert' and communes in 'minium' red. In addition to colour, style mattered. The army asked that its cartographers employ these tints lightly and required that, when drawing them on the sheets, the cartographer 'doit avoir deux pinceaux; celui qui fera le mieux le point servira à établir les teintes.' Furthermore, 'le papier que l'on choisira pour faire un dessin doit être bien collé et bien tendu sur une planchette ou sur un carton' (9–10). These precise details helped maintain a consistent level of quality.[21] Each sheet, although published separately, would appear the same way; a homogeneity the Cassini map lacked.[22] Hennequin

19 In one such example, Hennequin explains to readers that 'si on veut se servir de la carte pendant une promenade, le moyen le plus pratique, et celui avec lequel on ne se trompera jamais, c'est de plier sa carte en trois parties sur elle-même: on aura toujours la partie de la feuille sur laquelle on étudie à sa disposition' (22).

20 Hennequin labels 'la planimétrie' (the measurement of space) and 'le nivellement' (levelling) as the two most important components (7).

21 Hennequin provides additional tips for the Dépôt's cartographers. For example, the army states that 'les personnes qui s'exerceront à faire un lavis à l'encre de Chine procéderont avec des teintes excessivement pales qu'elles superposeront les unes sur les autres. C'est le moyen d'avoir des tons nourris et veloutés' (9–10).

22 While the Cassinis did give some suggestions on how to portray land on their maps, there was no specific colour for rivers. Pelletier explains that the Cassinis gave guidelines instead. For example, when it came to drawing

sheds light on the army's cartographic production and decision-making process as well as providing insight into the subtle ranking system it employed, for example in its portrayal of borders, and what the Dépôt may or may not consider cartographically important.

Surveying the Nation

With a sizeable corps of better-trained engineers by the end of the nineteenth century, France could rectify Cassini's personnel shortcomings. As demonstrated by the team of surveyors in the Alps, the recent changes and reforms provided the army's improved, more professional, and dedicated staff the necessary manpower to cover French territory. Originally the army asked its *ingénieurs géographes* to create a national survey very similar in size and scale to Cassini's multisheet work, the standard-bearer for French cartography since the eighteenth century. However, its workforce could now cover more terrain and gather intelligence more quickly. It became difficult to try to incorporate so much new information on map sheets of the same size used by Cassini and the first drafts came out overcrowded and unusable. General Demarçay worried that the Dépôt de la Guerre had strayed too far from its original mission to make French territory more accessible and comprehensible to the French public. Furthermore, it was not getting the results he expected.

> Comme on a mis beaucoup plus de détails dans la nouvelle carte, et probablement plus d'exactitude, mais comme les feuilles sont sur une petite échelle, comparativement à ce qu'on voulait faire en premier lieu, j'affirme que pour l'usage du commun des hommes, pour l'usage même des gens instruits qui ne veulent pas pâlir sur une carte, c'est-à-dire passer des heures entières pour étudier un très petit espace, je soutiens que la manière dont la carte de Cassini est gravée est plus satisfaisante que la manière dont l'est la nouvelle carte. (Berthaut I 265)

mountains, the Cassinis told engravers that 'à moins qu'elles ne soient à pic, elles ne paraissent jamais trancher durement, mais leurs sommets ont l'air au contraire de toujours s'arrondir, en s'allongeant plus ou moins, et former du côté de l'ombre un ton moelleux sans crudité' (140).

Demarçay argued to end the project. In 1832, frustrated by the lack of progress, he complained that 'On a dit qu'elle sera achevée dans 16 ans, si l'on remplit les conditions qu'a indiquées M. le Ministre de la Guerre; mais sur 238 feuilles dont elle doit se composer, il n'en a pas encore paru une seule, pas une seule! S'il y a quelques-unes terminées, elles n'ont pas encore paru.' (Berthaut I 265). The Dépôt had planned to sell the map sheets to the general public, but in their current state Demarçay doubted they would find much success. Skilled army personnel may have been able to use the map in its current form, but for the vast majority of the French, who had little or no map-reading experience, such a crowded map would be too confusing and impractical. However, General Jean-Jacques Pelet, head of the *ingénieurs géographes*, defended the project. He acknowledged the production difficulties, but reassured sceptics that the mission was in capable hands. He named illustrator Laplace to preside over the Commission in charge of directing the map project (Berthaut I 265). Pelet prevailed and, with momentum on their side, the map endured.

The Dépôt had begun its initial surveys in 1818, but it took ten years before it engraved its first sheet (Hennequin 6). The slow pace did not raise too many concerns at the agency. Taking into account the numerous errors found on the Cassini map, Pelet erred on the side of precision, even as it forced the Dépôt to recommence operations multiple times (Berthaut I 266). Returning to existing surveys, verifying them, and correcting errors provided the army with an accurate document it could use. But in order to obtain those much-needed specific measurements, the Dépôt had to improve training and field experience for its new recruits. Some workers struggled at first: 'Les ingénieurs du cadastre n'ont pas été en mesure d'exécuter cette triangulation de troisième ordre qu'on leur demandait' (Blondel 9). This lack of knowledge added to the project's cost and set back the schedule. In addition to the initial drafts' confusion and overcrowding, the fieldwork itself was uneven. Dépôt engineers calculated survey coordinates only to find that their measurements came up short. Furthermore, the Corps underestimated the time needed to survey a set area of land and officials realized that the work would continue for longer than originally anticipated (Blondel 9). The first map sheets ('Paris,' 'Beauvais,' 'Meaux,' 'Melun,' 'Provins,' and 'les frontières du Nord') appeared in 1833 and went on sale to the general public at the price of 7 francs apiece. A commercial success, this first round brought in more than 7,000,000 francs to the French treasury, a sum large enough to assure at least the short-term financial feasibility of the *état-major* project (Berthaut I 266–67). This lucrative preliminary print run also confirmed the market for maps in France. By the Third Republic, *cartes d'état-major* could be found in schools and even homes, as seen in *Le Tour de la France par deux enfants*.[23]

Although the Dépôt initially had doubts about employing general army personnel to help survey the terrain. But the sizeable monies raised with the first sheets confirmed their worth to the organization. The initial commercial success also helped foster pride in the project. Subsequent print runs proved equally profitable and, with the exception of a few sheets from the Alps, the Dépôt began to publish the remaining portions of the *Carte d'état-major*. By 1866, it had finished its final surveys (Bigourdan 433).[24] The engineers' time and effort in the field had provided France with the means to replace the Cassini map and the nation received its up-to-date military map almost 60 years after Napoleon first proposed one. The *Carte d'état-major* documented French territory accurately and its sale to the public provided relatively open access to information which, due to factors such as education and availability, would have been much more difficult to access a generation or two previous.

The Creation of a New Map

With the industrialization and sizeable demographic shifts that took place throughout the century, urban France continued to develop, transform, and grow well after the initial release of the *Carte d'état-major* sheets. As France's development progressed quickly at times, the Dépôt had to decide how best to revise its map sheets to reflect the current state of the nation. In 1860, it formalized a revision process and obtained updates from departmental prefects, most of which focused almost exclusively on new infrastructural improvements. Once the Dépôt obtained a modification, it would correct the

23 As discussed in Chapter 1, André and Julien are able to cross into France thanks to a walled *carte d'état-major* they see in their benefactor's home.

24 The Dépôt de la Guerre finished the remaining Alps sheets four years later, in 1870 (Bigourdan 433).

outdated or inaccurate data on its own maps. But this course of action had drawbacks. First, its own trained field engineers made occasional errors. So, local bureaucrats with much less geographic instruction made even more mistakes and often sent incorrect or incomplete data to the capital – General Berthaut referred to one set of received data as 'mediocre' (136). Other military commanders felt the same. On March 1866, Commander Déguilly of the *état-major* wrote to the Ministre des Travaux publics regarding the latest sheets his office had received:

> Je vous ferai remarquer, Monsieur le Ministre et cher collègue, qu'il serait très important, dans l'intérêt des divers services publics, que les rectifications fussent faites le plus exactement possible. Malheureusement, il n'en est pas toujours ainsi, et sur une des feuilles de la carte de France qui a été revue dernièrement pour la seconde fois, le Dépôt de la Guerre a dû rétablir des traces qui avaient été changes par suite d'une première revision [*sic*] envoyée par le préfet du département. (Berthaut II 136)

Contrary to the original plan of having local officials perform most of the preliminary work, officers from the ponts et chaussées ended up sending their own people out into the field to survey and provide modifications to departmental prefects who in turn sent the new information back to Paris (Bigourdan 434). The multistep process was far from perfect and the army's revisions were effected in a very piecemeal and out-of-order manner. One department may have a much-needed updated survey while its neighbour was still relying on a decades-old outdated sheet.

By 1867, the Dépôt de la Guerre developed a more succinct plan of action to execute needed revisions more efficiently. Colonel Hecquard, Lieutenant-Colonel Borson, and Commandant Déguilly of the *état-major* presented General Jarras, Chef d'état-major général, with several propositions that the army could employ to update the map more effectively. To learn about the construction of new rail lines, for example, the Dépôt would bypass the departmental prefects and instead query the Ministère des Travaux publics directly. When officers in Paris discovered a discrepancy on the *état-major*, the Dépôt could go ahead and send its own engineers to scout and verify the actual situation on the ground, and rectify what officials in Paris had in their possession. More extensive than rail, the French road network represented a greater challenge for the army's mapping agency. The only viable option for the Dépôt to find and correct roadwork changes was to dispatch available officers to record any up gauging, extensions, or improvements if, or when, alerted to them by local administrators.

An effective national map needed to portray the nation's current infrastructure, development, and urban growth. But due to the extent of French nineteenth-century industrial expansion, the Dépôt struggled to keep its maps current. For example, in 1850 France had 3,000 kilometres of rail. Fifty years later, in 1900, it had over 45,000 kilometres (Duby 289). In addition, industrialization transformed small towns and villages into manufacturing and mining centres.[25] Since it originally relied on local or departmental officials to provide construction data on a timely basis, the government's map seldom reflected the actual state of the nation. By the start of the Third Republic, military officials recognized that the continued revision process of the existing *Carte d'état-major* failed to happen fast enough and began discussing how to replace the map, as it had become obsolete more quickly than the army had predicted. For example, some of the Dépôt de la Guerre's initial surveying, still found on sheets used in *préfectures* around France, had been conducted prior to the inauguration of the first 1827 Saint-Étienne–Andrézieux rail line. Furthermore, it published many of its subsequent map sheets before France had completed the remainder of its rail system, laid throughout the rest of the nineteenth century.

Ironically, this outdated condition did not only affect the French administration. Other governments both directly and indirectly felt the effects of France's inability to modernize and bring its cartography up to date. In 1875, the *Gazette de Darmstadt* newspaper published an update on the German army's revisions of the French Alsace-Lorraine *état-major* sheets, at that point in the possession of the fourteenth and fifteenth German corps. The *Gazette* described the lengthy process required to revise the documents to meet German standards so that they presented the current state of existing infrastructure, now decades out of date. Since its original publication, dirt roads in the area had become rail lines and much of the construction completed in the previous 50 years had yet to appear on the map at all. Surprised, the Germans

25 The population of Saint-Etienne, one such example, rose from 33,000 in 1831 to over 110,000 in the early 1880s (EHESS archives).

had not realized how obsolete the French *état-major* had become. In France, officials knew that the map's progress had always been limited by resources and manpower. When it came time to complete its 1869 Seine department update, for example, the French army could allocate only two officers to cover approximately 220 square kilometres (Berthaut II 141). If the *état-major* could not provide more inspectors to document one of the most densely inhabited departments and home to the nation's capital, then the peripheral and less-populated departments would surely fare much worse.

In addition to these infrastructural inaccuracies, the topographical portrayal of France no longer met currently accepted norms (see Chapter 2). In the eighteenth century, the Cassinis too struggled with how to document land effectively. One nineteenth-century criticism of the Cassini map had been its lack of effective three-dimensional representations. The *Carte d'état-major* perpetuated this shortcoming, as no 'acceptable' solution had been found. Vidal de la Blache, for one, appreciated the significant manpower that went into the Dépôt map, but held that France needed 'une représentation plus analytique et plus rigoureuse du relief' ('La Carte' 115). In order to defeat adversaries on the battlefield, the army needed accurate and usable documentation of the terrain that soldiers and commanders could decipher and employ. To improve the quality of French cartography and provide the nation with the kind of maps it needed, the geographer hoped that any new map would be able to represent elevation more realistically.

Colour

The Service géographique's Paris *feuille* XXIII_14 presents the nation's largest city and as such serves as a showcase for the new *Carte de France* project. Vidal de la Blache wrote that 'Quoi qu'il en soit, on peut dire dès à présent que cette représentation des environs de Paris sera pour la précision des levés, la richesse des détails et l'effet artistique, un des plus beaux travaux cartographiques qui existent.' ('La Carte' 118). At a time when geography and maps had begun to contribute to the republican narrative and were helping to define 'Frenchness' and identity, *feuille* XXIII_14 played an important role in further introducing both the capital city and state cartography to the French. Furthermore, the Service géographique's standardized colour use during the nineteenth century homogenized what the French map reader saw. Thanks to authors such as Hennequin, classroom cartography increased, and the next time the reader saw another sheet or map of a French territory, he would recognize the Service géographique's cartography.

Colour plays an outsized role in disseminating the cartographer's message. In 1889, for example, Alphand's colour endorsed his enhancements to the Second Empire city. On the 'Les opérations de voirie exécutées de 1871 à 1889,' those bright improvements to Paris stand out, invite the reader's attention, and assert both progress and tangible results in the outer *arrondissements*. The selection of non-natural reds and yellows followed no predetermined formula or guidelines but allowed Alphand to champion his Parisian transformations over Haussmann's in the Second Empire.

Instead of following Alphand's method, where colour highlighted recent changes and updates to the city streets, in 1906 the army used tints to depict the status quo. The first colour that the user of the *Carte de France* noticed was certainly the red. Historically significant in French urban cartographic representation, this colour's presence here is no surprise as the army had used it extensively on its *Cartes d'état-major* throughout the nineteenth century. Similar to seventeenth- and eighteenth-century map-makers, who employed a red-like carmin to represent cities, readers familiar with this French historical norm would have easily identified the urban footprint on the Paris *feuille*.[26] Even though technology had freed map publishers from many conventional cartographic practices, as demonstrated by Alphand's bright yellows and reds, the Service géographique does not stray from its own standard. Hennequin's text had reminded readers that the army used red for 'Villes, villages, maisons' (8). The cartographer Pinon counted between ten and 12 other colours on the *Carte de France* (116). This chromatic range allows for a more natural or realistic depiction of the city and environs and takes into account land and water along with both private and public space. The multiple greens, reds, blues, and black correspond to different levels of vegetation, water, and urban development, and appear in all corners of Greater Paris. Red identifies the greater metropolitan area and the proliferation of that colour over such a large territory confirms the extent to which the city's suburbs

26 For a discussion of the importance of red in the history of French cartography, please see Chapter 1.

had spread since Alphand's 1889 portrayal. However, that same expansion also minimizes Paris's cartographic influence. Numerous towns and villages occupy land well beyond the city limits and Paris must compete for the reader's eye.

Power and influence conveyed through size, an effective tool to demonstrate worth, continued into the twentieth century with the army's *Carte d'état-major.* The Service géographique applied the largest, most visible font to two historically important cities: Versailles and Saint-Denis. From 1682, the year Louis XIV decamped there, until the 1789 Revolution, Versailles and state power had been interchangeable. Although the nineteenth century saw the government return to Paris, Versailles had served as a temporary headquarters to Thiers's Government of National Defence from 1870 to 1871 and to the Third Republic until 1879. Furthermore, as the *préfecture* of the Seine-et-Oise department, its role in regional administration justifies its identification with a large text. Saint-Denis, just north of Paris, is also clearly visible. This *commune*, at the intersection of the River Seine and the newly constructed Canal Saint-Denis (designated 'canal navigable') that bypassed Paris, was also historically prominent as home to the basilica that housed the remains of France's monarchs. By the turn of the century, industrialization also brought factories and workers to the northern suburb. The map clearly shows that development between Paris and Saint-Denis was erasing remaining vacant territory in between the two *communes*.

As the entity of its title and home to the infamous 'kilomètre zéro,' the point from which all road distances to the capital are measured, Paris occupies the centre of the map. The reader sees it clearly as the Service géographique employs colour to define the largest municipality on the sheet. Alphand had used Thiers's wall to define the limits of Paris in 1889, but by 1906 a pinkish tint now informs the reader where the city ends. The Service géographique showed that Paris's once-ultimate line of defence had finally been neutralized, not only by modern warfare – as military technology was capable of destroying Thiers's wall by the early twentieth century – but also by the increased commercial and industrial links between Paris and its suburbs. On the ground, the imposing barrier still marked a stark divide, yet shantytowns, factories, and suburban villages were beginning to blend city and suburb.

At the time of the map's publication, the government had yet to demolish Thiers's fortifications and would not be able to do so until after the First World War. But readers may have struggled to identify such an imposing physical border on the map. Perhaps ironically, the emptiness adjacent to the wall better partitioned Paris from suburban *communes* on this sheet than the actual barrier itself. The once-mandated 25-metre zone *non aedificandi*, open space reserved for military use, joins with the black *chemin de fer de ceinture* railways to create a visible divide between city and suburb. The mapping agency includes Thiers's wall on the XXIII_14 sheet, but small sections, near the Porte de Montreuil for example, appear to be missing. Elsewhere, development has pushed up to the barrier and camouflaged the wall's red outline. This growth downplays the wall's role and confirms that the fortifications no longer truly delimited the city. With stone and brick barriers now obsolete throughout Europe, cartography shows how vacant land could serve as the new divide on the *Carte de France*.

In the 36 years since France had last cleared the zone in preparation for the Franco-Prussian War, the reserve land had slowly vanished. Thanks in part to the large internal migration to the capital from rural France, housing shortages both facilitated and encouraged building there, most evident on this map near the Porte de Lilas, in Neuilly, Levallois-Perret, Clichy, and Boulogne-Billancourt.[27] In a few sectors, just outside of the twelfth, thirteenth, and fourteenth *arrondissements*, orchards and parks, both of which required human intervention, fill empty lots. The existence of maintained green space in the zone suggests that restrictions on new housing or development on the city's edge have been ignored or too difficult to enforce.

For that edge, the army presents commune limits in 'minium,' a reddish lead oxide (Hennequin 9). In most parts of the country this presentation choice, a thin line made up of a string of connected dots in minium red to mark the administrative periphery of a town or city, suffices. In the dense urban fabric of the Paris region, however, it falls short. Between the towns of Alfortville and Maisons-Alfort, for instance, the dotted line parallels a thick black rail line originating from the Gare de Lyon, which effectively neutralizes any cartographic benefit that it may have for jurisdictional purposes on the sheet. The border between Issy-les-Moulineaux and Vanves, another example, blends into the streets and buildings that compose the two towns, also negating

27 For a better description of the state of the fortifications between 1906 and 1919, see Chapter 5.

its employment. For Paris itself, the Service géographique marks not its communal edge, but instead the city's 'limite d'Arrondissement.' That perimeter limit – a plus sign followed by two dots – affirms that the capital city required a different jurisdictional border from other *communes* here. The *arrondissement* limit on the exterior of the vacant zone marks the administrative edge of the city. Contrary to the boundary of Issy-les-Moulineaux, which a reader would have to seek among crowded infrastructure, that particular line is easier to discern on this crowded map.

The cartographic blend of urban and suburban *communes* further emphasizes the minimized role which jurisdictional boundaries play in Greater Paris. As towns grow and fill in adjoining empty space, their individuality disappears on the map. In *mairies*, mayors held on to their local sovereignty, but on this sheet any significant differentiation between two separate towns, especially those close to Paris, appears minimal at best. The breadth of development across the map justified the more prominent role of combined regional urban sprawl on the map than any single *commune* on its own, including Paris. Furthermore, *feuille* XXIII_14 suggests that the government no longer viewed Paris as just its 20 *arrondissements* themselves, but as both the city and its expansive ring of industrial suburbs. Upon viewing the Service géographique's *feuille*, a provincial Frenchman may have agreed.

As the regional urban footprint spread, once-rural villages turned into suburbs and became integrated into the Greater Paris economic region throughout the late nineteenth century. The north-western suburb of Argenteuil epitomizes this extension well. The Seine-et-Oise *commune*, one of many outlying villages on the edge of the sprawl, sits on the banks of the Seine. Surrounded by pink vineyards, the Service géographique shows that it remained a semi-rural enclave. Yet the town terminates a line of development that follows rail lines departing from the Gare Saint-Lazare. Developers leaving and entering Paris would notice this open agricultural land and build on it in the next few decades. Other *communes*, such as Montreuil to the east of the twentieth *arrondissement*, Issy-les-Moulineaux just beyond the fifteenth, Ville d'Avray between Paris and Versailles, and Nanterre west of the Bois de Boulogne, presented similar undeveloped surroundings. With Argenteuil, they combine to delineate the western edges of Greater Paris. To the south-east, on the opposite side of the map, Créteil, Alfortville, and Champigny-sur-Marne all anchor a large accumulation of suburban *communes* around a bend in the River Marne. Vineyards, orchards, and fields here would befall the same fate as in the west.

From above, the combination of city, town, and village represents Greater Paris. At the street level, however, the Service géographique's individual buildings employ a darker shade of red that further defines the region and complements dense Parisian development. Historical, commercial, and industrial in nature, these structures form to create Paris's identity and contribute to the region's growing economy. Alone, the size of each square fails to accurately portray their impact. When pooled with the innumerable other red boxes throughout the enlarged metropolitan area, however, the clusters represent what makes the region work and affirm Paris's political and financial power. On their own, the squares are too small and numerous to label but indicate what citizens would see from the street as they moved through the city. A reader familiar with Paris would be able to identify many of the major historical and administrative sites so prominent in Alphand's 1889 atlas via outlines or adjacent landmarks: the Louvre, the Palais Royal, and Les Halles on the Right Bank, along with the Palais de Justice and Notre-Dame on the Île de la Cité. These structures had constituted a core of urban power for decades, if not centuries, and enhanced the concentration of dark red in the 1906 city centre so championed by earlier regimes.

Beyond bourgeois focal points, a Third Republic narrative of industrial and economic growth is also present. As the city's fiscal and demographic dynamics changed throughout the Second Empire and Third Republic, development migrated out from the historic centre. Many of the same administrative and manufacturing zones, prominent in 'Les opérations de voirie exécutées de 1871 à 1889,' exist on this XXIII_14 sheet too, and affirm that Paris retained its domestic manufacturing supremacy. For example, Alphand's infrastructural upgrades identified the growing economic influence of the warehouse district south of the Gare de Lyon in the twelfth *arrondissement*. By the time of this map's publication, the bright yellow maze of streets from 1889 had transitioned into one of the largest accumulations of dark-red boxes and rail lines, which drew the reader's attention away from the embellished administrative centre and testified to the economic weight and physical size of business and manufacturing in the city's outer quadrants. A second significant collection of industrial buildings breaches the northern border of the city, just past the abattoir complex so prominent on Alphand's 1889 cartographic document. This grouping starts adjacent to the *Marché aux bestiaux* and terminates in the suburb of Aubervilliers, just outside the wall. The dark-red arc validates the nascent influence of Paris's inner northern suburbs almost two decades before they acquired the *ceinture rouge* label following the 1924 parliamentary elections. In addition to identifying suburban power, this large cluster further confirms the decreased function of Thiers's walled defensive barrier. As the city's population, business, and industry expanded beyond the jurisdictional and physical city limits of Paris and more of these zones would cross over into, or appear in, the suburbs.

In addition to the sheet's dominance by red, the Service géographique's recording of significant amounts of forest and parks beyond the wall demonstrates the importance of green space to both the government and residents. Improved access to, and greater reach of, transportation no longer limited Parisians to the Bois de Boulogne and the Bois de Vincennes when they wished to escape the crowded city. The multiple shades of green outside Paris on the map document undeveloped, agricultural, and recreational sites, all of which complement the role of Haussmann's *intra-muros* Parisian city parks. The Bois de Boulogne and Bois de Vincennes, the pride and joy of Second Empire Paris, continued to serve the city's residents and the Service géographique affirmed both locations' value by assigning them the largest all-capitalized font on the map. The Bois de Vincennes, a mixture of woods and meadows and fields, still anchors the east side of Paris, but its composition lightens the park's cartographic footprint and renders it hard to see. The Bois de Boulogne, by contrast, contains more *bois*. The subsequent dark green (along with 'Prés, Vergers/Jardins' and 'Vignes') created with a mixture of blue and yellow 'gomme gutte mêlée avec une faible partie d'indigo,' gives the western park more cartographic visibility than its cross-town rival in the region's network of parks and forests. Unlike the 1889 map, where the two had an almost complete monopoly on the colour green, the Bois de Boulogne and Bois de Vincennes must now compete for the reader's gaze with other surrounding wooded areas, from the small Bois des Champious near Argenteuil, the Bois de Fausses-Reposes east of Versailles, the large Bois de Meudon, south of the *commune* of the same name, and the Parc de Saint-Cloud, home to the former royal residence. This collection of woodlands and recreational areas in the western environs of Paris reveals more open space on this side of the city than to the east, where factories and working-class shantytowns often occupy vacant land.

The Service géographique portrays the lighter *pré*, often adjacent to many of these forests, with 'indigo mêlé avec une faible partie de gomme gutte' (Hennequin 8). This complementary colour choice joins with green forests to create an illusion of an even larger expansive network of sanctuaries not far from the dense city. Due to lack of direct access to many of them, however, few Parisians or suburbanites were able to take advantage of this adjacent openness. Of course, city residents still had numerous parks inside the city to enjoy, such as the Jardins des Tuileries, the Parc Monceau, and the Buttes-Chaumont and the Service géographique documents them accurately. But the army's colour choice, combined with their small size, downplays these pre-Third Republic improvements. Outside the wall, woods, forests, and wild spaces, unconstrained by formal borders and the city, have not been divided or developed by man. *Feuille* XXIII_14's large surface area, combined with a significant amount of *pré*-like regional green locales, renders the orderly *intra-muros* parks less visible than sizeable woods and *jardins* outside the wall. This dichotomy depicts both a larger diversity and greater accuracy of wooded green space both inside and outside the city. It also suggests open land where, if needed, future urban growth can take place as more people moved to the region, further validation of the *Carte de France*'s potential.

Beyond the red city and green parks, the Service géographique employed a vibrant blue for rivers on the *Carte de France.* Water has always played a decisive role in Paris's development, for transportation and as a source of drinking water. In dark 'bleu de Prusse,' the Seine is visible and, in a significant difference from Alphand's 1889 map, is no longer the map's sole natural waterway. As suburban development expanded, other rivers played a critical role in the region's development. The Marne, adjacent to Alfortville and Ivry-sur-Seine in the south-east corner of the map, near the Fort de Charenton and the southern edge of the Bois de Vincennes, feeds into the Seine. In addition to the Marne, various *rivières* and *ruisseaux* 'important[s],' such as the Croult between Saint-Denis and Le Bourget, and the Morée between Aulnay-sous-Bois and Le Blanc-Mesnil, can be found on the north-east corner of the map. Local accessible rivers alone had never been able to satisfy the capital's thirst. Providing Paris with a reliable source of potable water has been an issue for French leaders for centuries. Napoleon's construction of a canal linking the River Ourq to the Seine via the Canal Saint-Martin, the last large-scale engineering project to address this necessity, demonstrates the lengths to which officials needed to go in order to provide this essential commodity to the capital's residents. The colour scheme highlights the former emperor's straight canal on the plain background and directs it from the north-eastern corner of the map into the northern sector of the city. The ability of the city's leaders to enhance its water accessibility by guiding these rivers and streams demonstrates the republic's ability to control nature for its benefit.

The Service géographique's more varied, tinted portrayal of Paris was significantly different from Alphand's for the many reasons discussed above. However, one noteworthy similarity remains. The army followed Alphand's lead in its portrayal of non-developed surfaces. In 1906, any 'vacant' land in the metropolitan area received no colour. The army had left any unoccupied or undeveloped territory the same colour as the background itself. This choice, even if made by default, confirms to the reader that the army charted exactly what existed: buildings, industry, infrastructure, and nature.

Topography

The Service géographique wished to map France in a manner that accurately represented the nation and at the same time helped the French population to understand it. To that end, the agency used associative colours: green for parks and blue for rivers. Furthermore, social and educational factors, such as the introduction and promotion of physical geography in schools and universities, allowed the Service géographique to improve upon its presentation of the nation's terrain as well. Topographically, the relatively flat Paris Basin did not pose much of a challenge for cartographers. However, many regions in the rest of France present with mountainous terrain that requires more map space and attention. In order to ensure that it did not overburden any particular sheet of the *Carte de France* with too many lines or symbols when demarcating elevation, the Service géographique reconsidered its reliance on the use of *hachures*.

In his *La Topographie mise à la portée*, Hennequin discussed the different means employed to portray elevation cartographically throughout the nineteenth century. Initially, the Dépôt de la Guerre had utilized both hash marks 'qui coupent, toujours à angle droit et qui indiquent les directions des pentes' and *courbes de niveau*, 'qui indiquent la coupe du terrain, par des plans horizontaux également espacés entre eux' (10). Hennequin admitted that of the two methods on the *état-major*, hash marks to depict elevation are the hardest to read (11). As early as the 1850s, cartographers realized that the *hachure* method obfuscated the actual terrain, took up too much room, or was just too hard for readers to decipher:

> On disait que la représentation par courbes ne permettait plus de rendre le relief du terrain saillant à l'œil, comme dans un dessin pittoresque à l'aide d'une lumière inclinée qui met dans la clarté un côté des montagnes et laisse l'autre dans l'obscurité. On reprochait aux partisans de cette idée d'adopter le système qui consiste, en somme, à couvrir les pentes d'une teinte (de quelque manière qu'elle soit produite, par hachures, par courges, au pinceau) d'autant plus foncée que la pente est plus rapide; en un mot, de quitter le système français pour le system allemand. On leur annonçait que les parties montagneuses deviendraient des plaques de noir où nul détail ne serait lisible. (Blondel 21–22)

For cartographers such as Blondel and Vidal, who also agreed with Hennequin, hash marks did not display territorial relief with enough precision and non-professionals had little use for the overly crowded terrain. Other methods showed promise. For example, the Service géographique had previously tested shading, where numerous small points form a continuous tint, which could then shade terrain to identify elevation. Many cartographers liked this method. Vidal, for one, remarked how this practice made dense terrain, often filled with infrastructure and elevation, appear 'more clear.' He mentioned specifically the sheets that presented 'des massifs sableux des forêts de l'Isle-Adam' and that of 'Carnelle' as good examples using the technique to their advantage. Closer to Paris, shading also highlighted the limestone embankments along the right back of the River Oise ('La Carte' 118). The larger debate on the best way to present three dimensions on a two-dimensional surface continued throughout the nineteenth century:

> Ici s'élève une question qui a longtemps divisé les écoles topographiques. [...] Comment doit-on représenter, peindre,

> pour ainsi dire, aux yeux, les formes de terrain? Tout le monde connaît le procédé suivi dans les cartes géographiques et ces petits amas de hachures, diversement groupées, qui marquent la place des chaines de montagnes; signe purement conventionnel qui équivaut exactement à cette phrase écrite: 'ici, il y a des montagnes.' (Blondel 19–20)

The Service géographique eventually settled on the *courbes de niveau* method and Hennequin describes and displays their illustrations in his *La Topographie mise à la portée* so that future cartographers would know and understand them. Lines, as opposed to hash marks, reflected an up-to-date cartographic norm and clarified altitude more easily for readers. The ten-metre *courbes de niveau*, or elevation lines, occupy less room than bulky hash marks on these maps. In addition to the standard ones, the army employed what Vidal refers to as *courbes de maîtresses*, larger 'master' elevation marks with a distance of 50 metres between them. For flatter terrain, which tends to house more development, the separation between lines leaves ample space for buildings and roads. On more hilly sections, multiple parallel *courbes de niveau* show verticality but still provide enough room for a map-maker to provide sufficient detail. Hash marks, on the contrary, generally take up too much map space and either hide or compete with map text and symbols.

As the nation's mapping agency worked to reform its cartographic depiction of land, France began to catch up to its European neighbours who had already made advances in their own domestic topographical depictions. De Margerie singles out Switzerland, home to much of the Alps, as a model map-producing nation that France should emulate. The contrast between maps of Berne, Zurich, or Brugg, which displayed considerable geomorphologic knowledge, and a sheet of the French *Carte d'état-major* of Lyon or Chambéry demonstrated how much France lagged behind. Hash marks on the *état-major*'s *feuille* 169 CHAMBÉRY depict the rugged terrain to the south of the Alpine *commune* and osbcure surrounding village names, such as Jacob-Bellecombette or Cognin. Sheet 169 shows the drawbacks to documenting terrain in this manner. De Margerie challenged geographers to place France's older *état-major feuilles* of the Jura Mountains alongside the 'admirable fragments' produced by their Swiss counterparts at the Commission géologique suisse. Other alpine countries had also made progress. Representation of volcanic terrain by Italy's R. Comitato Geologico map service further confirmed French cartographic inferiority (de Margerie, *La nouvelle carte* 242).[28] A transition away from hash marks would allow France to portray its own land more professionally, improve accuracy, and help its citizens better understand their nation.

Courbes de niveau streamline elevation markings and allow map-makers to include more infrastructure. On the XXIII_14 *feuille*, the 'full' pink city demonstrates why the technique is a better choice than hash marks. The Buttes-aux-Cailles neighbourhood in the city's south-eastern thirteenth *arrondissement*, shows how the Service géographique's demarcation lines record elevation effectively on a busy map. Several bands near the Place d'Italie reveal the area's steepness. Yet even with multiple lines present, the geographical agency was still able to incorporate all neighbouring streets and buildings.

Montmartre, on the city's northern edge, provides an even better example of how to effectively portray a cityscape in an exceptionally vertical environment. As it sits high above the city, *courbes de niveau* abut the Sacré-Cœur basilica and elevate it cartographically. The large number on the plateau's southern edge do not obscure that part of the map. On the ground, the area's verticality hinders much actual building, so that the proximity of the lines does not conceal any information. Their presence further confirms the sharp incline between the basilica itself and the future green park below. A few other sites also demonstrate the efficacy of documenting elevation in the dense urban fabric. In the Passy section of Paris's sixteenth *arrondissement*, two long *courbes de niveau* portray the Right Bank's rise from the Seine as the *ligne* 6 of the Paris *Métro* heads below ground on the way through the Trocadéro towards the Place de l'Étoile.

The general shape of the Paris Basin appears on the outskirts of the XXIII_14 *feuille*, primarily with lines in the north-eastern corner, which presents hills past Le Blanc-Mesnil. To the east of the city, close lines near the *commune* of Fontenay-sous-Bois indicate the vertical disparity between the Fort de Nogent to its south and the Fort de Rosny to its north. To Paris's south, the town of Fontenay-aux-Roses sits within a line of hills. In this sector of the map, the reader can follow the *courbes de niveau* from Bagineux to its north, around Fontenay-aux-Roses to the

28 The *Carta geologica della Sicilia* (1882), one such example, offers a multicoloured depiction of the island, including Mount Etna.

west as they approach Fort de Châtillon and continue on to the *commune* of Plessis-Fontenay. Finally, adjacent to the Bois de Meudon to Paris's south-west, another significant band of elevation differentials runs through the green forest from north to south. Here, the forest's location demarcates a break in development and suggests that its relatively rural location and steep elevation have not deterred development.

Overall, *courbes de niveau* accurately represent the region's elevation differentials and allow the Service géographique to map infrastructure, nature, hills, and valleys. Their use and success, however, relies on reader knowledge of them. Hennequin's text explains elevation lines and includes images of them. The army's mapping agency, on the contrary, did not include any graphic representation of the *courbes* in the *Carte de France* legend. Instead, map users had to read the description found there:

> Le nivellement de la feuille a été rapporté aux cotes de nivellement Bourdalouë: l'équidistance des courbes est de 10 mètres; inscrites isolément ont été arrondies au mètre. La constante régionale de correction pour rapporter le nivellement de la feuille au zéro normal du nivellement général de la France est_0m.

This choice also confirms the sheets' target market: the educated bourgeoisie. The complete lack of explanation of symbols in the legend precludes the illiterate or poorly educated from understanding. Of course, the Service géographique hoped that such people would still purchase the map, even if unable to use it to its full potential. Most likely, they would put the map sheet to a more decorative use, similar to André and Julien's guide in Fouillée's *Le Tour de France*.

Legend

To ensure comprehension of the 1906 *feuille* XXIII_14 of Paris, as in the earlier *Carte d'état-major*, the military mapping agency had to explain its colours and symbols, and the legend to the *Carte de France* provides a key to the 23 symbols and the colours and lines used on the map. The legend thus plays a crucial role in map reading. It educates the reader and defines what he or she sees on the map. Remi Caron states that:

> Il faut d'abord comprendre comment fonctionne la légende. En théorie elle consiste à doubler chacune des relations entre

> les objets et leur représentation cartographique par une autre relation entre ce signe cartographique et un mot. Autrement dit, pour expliquer au lecteur le lien entre un objet du monde et son image sur la carte, on fournit d'abord l'expression, en langue française par exemple, de cet objet, puis on reproduit en face de ce mot le signe graphique correspondant. C'est donc bien une traduction, qui a de particulier qu'elle cherche à superposer une langue à un langage de signes graphiques. ('La Légende' 329)

As a translation, a representation, or an expression, the legend takes on many roles, yet first and foremost, it presents the cartographer's point of view. On the *Carte de France*, the legend demonstrates which elements the Service géographique wished to prioritize. On this map, it is the growing suburban *communes*, accompanying bureaucratic and industrial infrastructure, nature, and terrain. The scale of the territory which the Paris sheet covers meant that individual streets could not be named and all but required the Service géographique to employ signs to document the region's growing, diverse urban footprint.

The Corps des ponts et chaussées had spent much of the nineteenth century building and expanding France's route network, with Paris being the terminus of main highways and railways, linking it to the rest of the nation. In recognition of its sister agency's efforts, the Service géographique's cartographers placed the road infrastructure list on the dominant left-hand side of the legend. The army's map validates the comprehensive status of French transportation in the capital's region and shows in-depth knowledge of its infrastructure. At the top of that list, the reader will first notice 'Routes et Chemins Carrosables régulièrement entretenus.' Descending this first column, one finds 'Route Nationale,' 'Route Départmentale et Chemin de Grande Communication,' 'Chemin ayant au moins 6m,' and 'Chemin ayant moins de 6m.' Following, the Service géographique displays 'Routes et Chemins Carrosables Irrégulièrement entretenus (Chemins de toutes largeurs).' The quality and significance of roads diminishes as the colour of the line fades and grows thinner. The *routes nationales* are darker and therefore more noticeable and more important than departmental road in the eyes of the army and of potential customers or users of the map. Beyond interpreting roads, the Service géographique's inclusivity also incorporates all levels of pathway. At the bottom of the list, for example, the reader can find the category 'Chemin d'exploitation, Sentier de piétons permanent, [et] Vestiges de voie ancienne,' which demonstrates how detailed military surveying had become and affirms the Third Republic's mission to know and document every road, street, and footpath in the capital region.

The Service géographique's *Carte de France* displays rail's influence and accessibility in the capital region in a similar manner. The solid, dark double lines of the 'chemin de fer à quatre voies' appear more prominently than both the walled fortifications and most developed roads in the region. The Service géographique's presentation of four-lane rail lines differs only slightly from the hollow *route nationale*. The Gare du Nord, for example, sends one of the largest, most prominent lines on the map northward towards Chantilly. Similar to the road network, the different levels of train infrastructure illustrate comprehensively the state of French light and heavy rail. Access to public and private transportation provided Parisians of all income levels the means to circulate inside and outside the city. The tramways and omnibus lines that carried Parisians to parks and across the city in 1889 now join rail to offer those same residents the ability to venture out further from the city centre.

The recently inaugurated underground *métropolitain* inside the city, although completely independent of suburban and intercity rail companies, complemented the national domestic rail network as it helped travellers move between neighbourhoods and the city centre. The Service géographique, no doubt happy to showcase the French engineering that built the underground train system, documents both above ground 'en viaduc,' with solid lines, and below ground 'souterrain' subway lines, with dotted ones. *Métro* and tramway routes crisscross the city, occupy many of the wide boulevards, and assume the role that Haussmann's *grands boulevards* once held as they connect travellers to train stations serviced by 'lignes à quatre voies' in all corners of Paris. Transportation's position in the Service géographique's legend reaffirms what French leaders had believed for centuries, that moving people through the crowded city, and now wider region, was good for residents, for the government, and for business. Under the Second Empire, Haussmann started this process on a large scale. He opened up the congested centre and demolished numerous medieval *îlots* in many *arrondissements* to build wide boulevards, and the Third Republic continued this work. With an influx of new residents, the city and environs required more roads and railways, and the *Carte de France* demonstrates the government's work to address this need.

As it did for roads and rail, the army also graded waterways, though here it ranked not by colour, as all water appears in the same 'bleu de Prusse' tint, but by width. The Seine has traditionally been one of the city's principal 'thoroughfares.' The agency labelled that chief waterway, among others, 'Fleuve ou grande rivière.' The next four levels in prominence are the 'Rivière, ruisseau important,' the 'Canal navigable, rivière canalisée,' and finally the 'Source, fontaine, puits.' Through this documentation, the army asserted its knowledge of, and control over, the capital region's navigable rivers and streams. In addition, it included waterway improvements: 'canal navigable, rivière canalisée' and 'Port, écluse et maison d'éclusier,' along with bridges that cross these rivers. The Service géographique's river diagram, traversed by different standards of bridge, would be of obvious interest to the engineers working for the Ponts et chaussées agency and its own *ingénieurs-géographes* who surveyed the national terrain. By the start of the twentieth century, 24 'Ponts en pierre, en fer, en bois, [et] suspendu' spanned the Seine inside the city limits and an additional 27 did so beyond them. Between floods and droughts, the Seine had been a natural obstacle for man to master since Roman times. Henri IV's Pont Neuf, the first permanent stone bridge to straddle right and left banks, transformed transportation in the Renaissance city. Since that first direct traverse, subsequent rulers also looked to improve river crossings in order to enhance movement in the crowded city.

To the right of water in the legend, the Service géographique also showed the importance of French industrial production. While the massive red sections of the mapped city proved that much of Paris and the surrounding area accommodated factories, the legend provides a more detailed diagnosis of the state of French industry at the start of the twentieth century. The 'Etablissement industriel' subsection is divided between those factories that are 'mû[s] par la vapeur, à moteur hydraulique, [et] producteur[s] d'électricité.' With the exception of the hydraulic station, symbolized by a blue turbine wheel, the industrial sites on the map are presented as red miniature factory buildings. This colour choice, however, blends them into the rest of the regional infrastructure and makes them hard to find. The dirty factories provide jobs for newly arrived residents but the Service géographique felt no need to highlight them. From north-east to south-west, from La Villette to Saint-Denis to Boulogne-Billancourt, the map reader sees much dark red and the crowded industrial band hosts many miniature factory-shaped symbols. Though many others, adjacent to the Seine in Puteaux and Courbevoie, are easier to identify and foretell the rise of industry in these *communes* further west.

Other categories in the legend further document the scope of the Republic's reach. With the exception of recent World's Fairs, Third Republic France, unlike other nations and regimes, did not favour construction of massive public works to promote itself. French kings and emperors had been much more active in their physical embellishments. Instead, by the end of the nineteenth century, history and identity, through the creation and promotion of Republican symbols, helped citizens understand and endorse the nation. The powerful symbols included the tricolour flag, the Republican monogram RF, the motto *liberté*, *égalité*, *fraternité*, and the national anthem, the *Marseillaise*, which hailed back to the Revolution. 'There was to be no official national day other than 14 July, no formal mobilizations, processions and marches of the civilian citizenry, but rather a simple "republicanisation" of the accepted pomp of state power – uniforms, parades, bands, flags, and the like' (Hobsbawm, *Invention* 271). Republican symbols epitomized the spirit of the regime. If needed, a physical presence could reassure the wavering citizen.

The limited surface area on any map, regardless of size or scale, constrains the cartographer's ability to include as many features as he or she may like. In general, map-makers need to accommodate a considerable amount of detail in a very small space. Therefore, symbols allow the unambiguous presentation of pertinent information to as many readers as possible (Caron, 'La Légende' 329). The presence of icons on the *Carte de France* also corresponds to an increase of their use in general in Belle Époque society and in French political discourse more specifically.

To this end, the Service géographique employed an envelope on the map for the 'Bureau[x] de Poste,' an envelope and a wire antenna for the 'Bureau[x] de Poste et de Télégraphe,' and an envelope and a larger wire antenna for the 'Bureau[x] de Poste et de Télégraphe et Téléphone.' This communications enterprise appears extensively on the map, though curiously only outside the city boundary. This void suggests that Parisians knew where to find postal facilities and that their ubiquitous nature made it both impractical and unnecessary to include them there. Outside the city, however, one can spot *bureaux de poste* and *bureaux de poste et télégraphe* near *routes nationales* or in the centre of *communes*. The *bureau de poste et de télégraphe et téléphone* represents a functioning state agency whose quick and efficient dissemination of information affirmed how much the Republic worked for its citizens. For example, the *facteur* played an important role in daily life. The well-liked civil servant not only distributed mail, but also sold residents stamps, helped them open savings and checking accounts, and distributed news and events (Richez 39–40). Furthermore, the postal service provided well-paid jobs to France's newly educated population. According to the map, residents could now find where to send and receive post and telegrams. In addition to the main postal branches, the map includes many *bureaux télégraphiques ou téléphonique*, both 'public[s]' and 'non ouvert[s] au public.' These buildings are smaller than post offices, which perhaps reflects a less frequent interaction by the average citizen.

As it documents the expanding communication network, security, and healthcare throughout the entire Parisian Basin, the *Carte de France* presents an optimistic nation, where the government harnesses science and technology to surmount obstacles. The Service géographique's legend assigns a red triangle inside a black circle for 'Gendarmerie.' The two colours together help these police stations stand out on the predominately red document. Their presence in the city will also ensure stability. In a nod to international norms, the army employs a red cross in a black square, in use since its adoption in Geneva in 1863, for 'Hôpital.' 'Mairies' are represented by a red rectangle in a black box; a red circle indicates an 'église,' a red cross on a red box for 'chapelle isolée,' and a cross for 'croix.' In addition, the Service géographique rounded out its legend with 'four à chaux,' 'four à plâtre,' 'carrière à ciel ouvert,' and 'carrière souterrain.' There are numerous symbols found throughout the map, but the Service géographique believed that its readership could navigate them easily. The legend's presence, along with the large variety of colours and symbols throughout this mapped area, confirms an increased fluency and comfort with reading and using cartography at the turn of the century.

Conclusion

French royal maps had existed for centuries. However, by the beginning of the Third Republic, national agencies began successfully employing cartography to document the city space and, perhaps more importantly, endorse the government. As France entered the twentieth century, a burgeoning middle class, coupled with greater economic opportunities and participation in the governing process, resulted in a wealthier, more self-assured population. Official representations of Greater Paris existed before *feuille* XXIII_14. The Cassinis' *feuille* '1: Paris,' published in 1749, focused entirely on the city and surrounding territory. The army's nineteenth-century *Carte d'état-major* sheet number 48 did the same. Later, Alphand's 'Les operations de voirie de 1871 à 1889,' part of a large atlas commemorating the 1889 centennial, shows that the government wished to document its capital. Citizens with access to these maps could find and view accurate portrayals of Paris.

In an effort to update Cassini's groundbreaking map, Napoleon first suggested a new cadastre map to help him administer his empire, although he failed to achieve this goal before his final defeat and exile. After the emperor's fall, the army took charge of creating an up-to-date national map. It dispatched surveyors throughout the nation who created a homogenized mapped portrayal of French. The resulting *Carte d'état-major* presented much of France's interior to politicians, military leaders, and the general public for the first time. This introduction coincided with the growth of the map reader, which provided the Dépôt de la Guerre and successor agency the Service géographique with a larger potential market for their products. The maps served other purposes beyond being purely informational. Establishing a republic after almost a century of regime changes – a *Directoire*, monarchies, and empires – the French state needed to prove to its citizens that it could administer the large nation effectively. Maps joined in this drive to provide citizens with a tangible piece of evidence that defined 'France' for much of the nineteenth century.

Republican politicians looked for additional ways to promote the Republic's message of championing unity and a common national identity. It relied on schoolbooks, maps, and even the army to legitimate itself to both its citizens and the wider world after its loss to the Germans in 1871. André and Julien's use of a government map to cross into France successfully demonstrates how improved and expanded map exposure would benefit the country and it became one effective medium through which to broadcast this message.

As map-printing technology advanced and rival forms of government no longer posed a threat, republican officials grew more confident in their ability to remain in power. At the same time, maps became more sophisticated in their promotion of the regime. Green parks and woods, for example, offered a nearby refuge from the chaotic city. Blue rivers suggest an abundance of clean drinking water. Furthermore, the idea of cities also started to change. Considerable growth all but eliminated any meaningful differentiation of where individual towns started and ended as they merged with the Parisian urban sprawl. When referencing 'Paris,' did one refer only to the land inside the wall, or both the city and its suburbs? The 1906 *Carte de France* suggests that the government were of the latter opinion and shows Greater Paris's shape and size. As one of the first sheets published, the map of Paris affirmed the capital's value to the nation.

For much of the nineteenth century, French cities represented a rebellious segment of the population. They brought down leaders and governments. Once calmed, however, urban areas provided benefits to France as well. Embellished city centres epitomized modernity and attracted visitors. Factories on the city's edge equalled jobs and higher income for French voters. In many respects, progress in Saint-Simonian terms meant that industrial, urban, and suburban Paris should grow even at the expense of open green space. For these reasons, the Service géographique's 1906 *Carte de France* presents a capital city that epitomizes what works in the Third Republic. Compared to Alphand's 'Les opérations de voirie de 1871 à 1889,' where the government is shown to have improved the city, here it is shown to be thriving, having expanded considerably thanks to republican stability.

Colour remained an important cartographical tool, dark reds highlighting this industrial, commercial, and extra-mural residential development. The addition of a legend validates the capacity and willingness of French map readers to read one and to use its symbols to decipher the agency's cartographic discourse. It also documents a more sophisticated cartographic relationship between reader and map-maker. Not only could a turn-of-the-century cartographer create a more complex map with colours and symbols, but he or she can also use it to promote France. Multitracked railroads, paved roads, canals, and bridges promoted France's engineering successes. The many factories also demonstrated France's ability to provide for itself and to export to neighbouring countries. Finally, post offices, police stations, hospitals, and telephone stations showed a nation with a functioning bureaucracy. France could heal, police, and communicate with its citizens.

Of course, a map such as the *Carte de France*'s *feuille* XXIII_14 could not be produced without state backing. Nonetheless, the commercial nature of this sheet, sold to the general public at large, broadened the project's scope and influence significantly. Considered up-to-date at the time of publication, by 1914, as war broke out, the *Carte de France* soon revealed its limitations, including 'son manque de précision [et] son absence de quadrillage kilométrique' (Pinon, *Les Plans de Paris* 116). These deficiencies eventually led to the publication of a less colourful, more topographic *état-major* map in time for the war's end. Yet, *feuille* XXIII_14 led the way to greater variety and specialization in cartography as the century progressed. So when technology changed how the French interact with urban space, personal cars for example, French cartography easily adapted.

Both the Dépôt de la Guerre and the Service géographique helped the nation's military better understand the topography of France, including Paris's environs, in order to construct, organize, and deploy an effective defence of the city. Yet perhaps the real power and importance of the *Carte d'état-major* and *Carte de France* lies in their role in helping the government to plan the city and nation's future, even if unofficially. The Service géographique's cartography served as a base for much of the urban planning in the first half of the twentieth century. Following the First World War, the government looked to manage and reform the growing French urban footprint and called on politicians and academics to provide solutions to the problematic crowding and congestion which the new influx of residents brought with them. Since legislation often required a cartographic visual of any proposed changes to accompany any plan, the fastest and easiest way for potential urban planners to produce a map was to employ the government's existing cartographic resources. When Léon Jaussely presented his *Projet lauréat de la section générale du concours du plan d'extension de Paris* (*Plan d'extension*) to Paris officials in 1919, for example, his multicoloured map showcased his ideas for the French capital. To create that plan, he simply painted these ideas on top of an existing Service géographique map. This was possible not only due to his faith and trust in the document itself, but to that of the officials receiving it. The amount of manpower and energy devoted to state mapping throughout the nineteenth century created a valuable resource for not only the government, but also the public at large. *Feuille* XXIII_14 showed millions of French people what their capital city looked like. The resulting knowledge will help the Third Republic govern effectively.

Chapter 5

The Beginning of French Urbanism

Léon Jaussely's 1919 *Plan d'extension*

> Pour la première fois une revue française ouvre une chronique de l'Urbanisme. *La Vie Urbaine* ne pouvait manquer de prendre cette initiative: créée par le Conseil municipal de Paris pour l'étude méthodique des phénomènes qui président à la formation originelle des agglomérations, puis à leur développement [...]. Dans l'ordre des recherches scientifiques, le champ de l'urbanisme est très vaste; n'en étant qu'au prémisses, cette science a devant elle un avenir étendu; dans l'ordre des réalisations pratiques où l'on cherche à appliquer les données de la science à la construction des agglomérations modernes, l'horizon, par le fait de la guerre, s'est singulièrement élargi.
>
> —Léon Jaussely, Architecte en Chef du Gouvernement, editorial for the inaugural edition of *Urbanisme* (1919)

As Paris entered the twentieth century, politicians and leaders from suburban communities increasingly articulated their dislike of the Paris-centric administration of the Seine department. Albert Thomas, former Undersecretary of War and mayor of Champigny from 1912 to 1919, perhaps best articulated the growing city-suburb divide. In his preface to a 1920 government report, Thomas spoke out against the economic segregation between the two sides. While the suburbs gained many new residents from rural France, they also accumulated costs associated with housing, caring for, and educating them. The factories, workshops, and stores that employed the newcomers, however, sent their taxes to Paris, where they usually remained (5). He called for more attention and development across the entire department, especially in suburban communities that experienced the most growth.

In many respects, Thomas was right. Between 1872 and 1911, the city and adjacent Seine department gained over a million new residents (Wolf 17). Neighbouring departments also grew considerably in the first four decades of the Third Republic. In the 1911 census, the Seine-et-Oise claimed over 818,000 residents and the once-rural neighbouring Seine-et-Marne to Paris's east was home to 364,000 people. Now suburban communities needed to cooperate with Paris, boxed in by an outdated wall, to look for efficient and effective ways to administer, control, and plan for the region.

As the Paris Basin become more crowded, changing ideas of urban environments, championed by the Musée social movement plus the creation of the Société française des Urbanistes along with the subsequent École des hautes études urbaines, altered how government officials viewed the region. Thanks to the influence of these institutions, the idea began to take hold more generally among both intellectuals and politicians that the French government should plan Parisian urban growth, not just react to new and existing problems. In addition, with accountability to voters and such a large area to govern, the French state's traditional top-down hierarchy, exemplified by a strong personality such as Haussmann or Alphand telling a community what it needed, could no longer continue. Until the twentieth century, the desire for a modern city and suburbs reigned supreme and officials championed aesthetics as much as improving movement or peoples' lives. Yet, thanks to innovations introduced by the Musée social, French urban reform would begin to focus attention on creating healthy and hygienic housing and raising living standards for the urban area's residents, many of whom lived in squalor and poverty outside the city limits.

Of course, not everyone agreed with the need to spend time and money improving the lives of the poor. Bourgeois Parisians compared themselves not with small ramshackle communities beyond the wall, but with other European capitals such as London and Berlin. In those metropolitan areas, along with many in the United States, a new science of cities called 'urbanism' gained in popularity. The discipline changed how people viewed and used urban space and greatly enhanced living conditions for residents. In France, however, the term 'urbanisme' itself arrived later. It did not even appear in the French language until early 1910, in a small study by Pierre Clerget, 'L'Urbanisme: l'étude historique, géographique et économique.' It became more widely used later that year with the creation of the Société française des urbanistes (Jaussely 187). How did this group apply the science of urbanism to French cities? The replacement of prestige projects with more green space, woods, and parks joined with transportation as priorities in subsequent French urban legislation and planning.

This chapter will examine the rise of French urbanism via the debate surrounding the demilitarization and dismantling of Thiers's outdated fortifications and the subsequent rise of the Musée social movement. These two developments influenced the urban legislation of the 1919 loi Cornudet considerably, Jaussely's ensuing Projet lauréat de la section générale du concours du plan d'extension de Paris map, as well as subsequent twentieth-century French urban plans. The desire to regulate and control suburban growth, and the eventual *Plan d'extension*, represent not concrete plans but optimistic possibilities. But, perhaps more important, taken together, both events mark the beginning of modern urban planning in France. Finally, these urban transformations occurred as technology changed the overall idea of cities and their accompanying urban space in France before and immediately following the First World War.

Jaussely's Map – *circulation, hygiène, esthetique*

Due to its size and importance to the nation and its economy, Paris remained distinct from most cities when it came to urban reform. In 1915 François-Joseph Cornudet des Chaumettes, a deputy from Seine-et-Oise, brought to the Chambre the first piece of legislation that would lead to the regional plan first proposed back in 1906 by Forestier. Cornudet's bill stated that 'Toute ville de 10.000 habitants et au-dessus, sans préjudice du plan général d'alignement et de nivellement imposé à toutes les communes par l'article 136, 13°, de la loi du 5 avril 1884, est tenue d'avoir un projet d'aménagement, d'embellissement et d'extension.' But communities in the Seine department with 'moins de 10.000 habitants et de plus de 5.000 dont la population a augmenté de plus de 10% dans l'intervalle de deux recensements' also needed a plan. Outside the Seine department and large cities, 'Stations balnéaires, maritimes, hydrominérales, climatiques, sportives et autres dont la population, quelle qu'en soit l'importance, augmente de 50% ou plus à certaines époques de l'année' also required a plan (actual legislation qtd in Remaury 227–28). Perhaps due to a desire to clean up slums, hinder working-class dissent, or just an impulse to meddle in the affairs of Paris, home to the Chambre des députés, the government focused first on the largest French cities, but also targeted those that housed large influxes of people seasonally, coincidently where many politicians and urban elites vacationed. In addition, the effects of the First World War loomed large. The damage from that conflict affected large sectors of Northern France, and so the final version of the loi Cornudet also required cities and towns partially or completely destroyed by war to create plans, with the government agreeing to pay the associated costs. First introduced in 1915, the final version of the bill did not become law until March 1919, after the war ended. Finally, if any urban plan crossed departmental borders, then an interdepartmental conference had to be created (Prost 228–29). Introduced in 1915, the bill did not become law until March 1919, after the war ended. As the legislation moved through parliament, the growing field of urbanists hoped to use these reforms to improve the lives of residents.

Léon Jaussely joined with Roger-Henri Expert and Henri Sellier to submit their Projet lauréat de la section générale du concours du plan d'extension de Paris, also called 'Urbs' to the Commission d'extension. In accordance with the parameters of the competition, each entrant was required to furnish a map of their chosen scale. However, the only condition was that entries 'devront être, groupés ou isolés, présentés sur des feuilles dont la dimension n'excédera pas 1 m. x 1 m 40. Les concurrents de la section I auront droit à quatre feuilles au maximum.' To provide cartographic uniformity, the Commission stated that applicants needed to employ black to represent 'les parties [de l'état actuel de la région] maintenues,' red for 'les parties à créer,' and yellow for 'les parties à supprimer.' For anything else he wished to include, 'le concurrent aura toute liberté de présentation' (101). This flexibility allowed Jaussely to adhere to accepted cartographic norms (green for parks, blue for water, etc.) but also employ non-natural tints to present a vibrant, modern Paris; a capital city that welcomed the future, took care of its citizens, and managed the region. The trio took home the *première prime* in the Section I category, reserved for those projects that incorporated maps and a regional plan as defined by the loi Cornudet.

The large *Plan d'extension de Paris* of Jaussely's 'Urbs' project promoted what he, Expert, and Sellier wished to see in a modern

French metropolis and provided a glimpse of the future. Equally important, it demonstrates influence from the French school of urbanism and the Musée social in its recommendations for cities to provide healthy living environments for residents and visitors. The Musée social gained much attention due to its participation in the dismantling of Thiers's wall. Since much of the debate on what to do with the fortifications had taken place around the 1908 municipal elections, almost a decade earlier, lawmakers and commission members wanted to see how planners envisioned using this space once it became available. Some parameters were already fixed. For example, Louis Dausset's fortification project, presented to the city of Paris in 1908, called for the construction of a limited number of *habitations à bon marché* (HBM) and military barracks along with the sale of some of the zone's most valuable land. However, in the majority of the zone to be vacated, the city wished to construct parks, playing fields, and a Palais des Expositions (already promised via an agreement with the Ministry of Agriculture) which would be located in a section of the wall between the Porte de Meudon and the Porte de Saint-Cloud ('Le Programme et le résultat du concours' 94). Jaussely's *Plan d'extension* allows for all of these previously negotiated additions.

To make any project successful and worth the Commission's time, authors needed to understand geography and to react appropriately to the region's terrain. Myriem Foncin, a *géographe bibliothécaire* who, in 1920, became the first female librarian in the department of Cartes et plans at the Bibliothèque Nationale and in 1942 the first female director of a department at the Bibliothèque Nationale, offered her opinions on many of the various proposals of the loi Cornudet in an essay, 'Quelques réflexions géographiques à propos du concours (1ère section) pour le Plan d'aménagement d'extension de Paris,' in the 1920 issue of *La Vie urbaine*. Foncin noted that many entrants believed that they could put roads and buildings anywhere, regardless of feasibility. For that reason, she praised Jaussely's 'Urbs' project for its exactness, describing it as 'Le mieux documenté et le plus complet' (79–80). Jaussely's familiarity with the terrain helped him create a practical and feasible plan for the city and his win provided him with worthwhile distinction.

Thanks to Vidal de la Blache, the growing role of physical geography in France shaped how intellectuals, such as Myriem Foncin, viewed the French soil and influenced how planners suggested future expansion. Foncin relates the actual terrain with Paris's growth:

> Si l'on regarde une carte de la région parisienne indiquant à la fois le relief et la répartition des habitations, (le nouveau 1/50 000ᵉ en couleurs par exemple) on s'aperçoit vite que l'une est fonction de l'autre. Le centre urbain proprement dit s'étale dans la grande dépression que la Seine a déblayée entre les buttes ou les plateaux de Montmartre, Belleville, Villejuif, et Chatillon. Dernièrement, il a un peu mordu par endroits sur ces parties hautes qui ont longtemps limité son développement. Mais il a surtout poussé des tentacules en bordure des rivières: le long de la Marne: le long de la Seine surtout vers l'amont [...] comme vers l'aval. [...] L'agglomération urbaine s'étend donc presque exclusivement dans les vallées. (84)

Reminiscent of Vidal's portrayal of the Paris Basin as the natural, geographic home of the French capital, Foncin commends those projects that also conform to the actual physical limits of expanding the Paris agglomeration.

For this reason, Jaussely's 'Urbs' plan was lauded by many intellectuals, including Foncin. A geographer by training, she faulted many of the other projects for their authors' blatant ignorance of the Seine department: 'Si presque tous les travaux envoyés au concours trahissent une grande inexpérience ou même une certaine ignorance géographique, on ne doit pas s'en étonner.' Most authors, she lamented, had little to no knowledge of physical geography, which was made obvious by the impracticality or even impossibility of their proposals. 'Projet Sirius,' one such example submitted by Emile Molinié, Charles Nicod, and Albert Pouthier, received a stunning rebuke:

> Ils ménagent d'heureuses perspectives, mais les routes et les lignes de chemin de fer nouvelles qu'ils prévoient passent indifféremment des vallées sur les plateaux, traversent des centres importants. Leurs canaux escaladent les hauteurs comme ce canal exutoire qui, quittant la Seine à Javal (26 m), monte en suivant les fortifications désaffectées sur la butte de Monsouris (76 m), descend rejoindre la Seine à Bercy (30 m), remonte sur la butte de Belleville (125 m) pour se terminer au basin de la Villette (50 m). (Foncin 78–79)

Moving water uphill and constructing rail and roads may look nice on paper, but it would be impossible to construct this infrastructure in real life. To be successful, planners had to know the region's terrain.

Delimiting the Region

Before Jaussely could attempt to plan Paris and its suburban belt, officials had to identify and demarcate the growing metropolitan area. How could officials delineate exactly which areas should be targeted? As demonstrated by the army's 1906 *Carte de France feuille* XXIII_14, Napoleon III's official expanded city limits of 1860 no longer reflected the true state of the capital. Thiers's fortified wall, a physical barrier designed to defend the city from external armies, now hindered development on the city's outskirts and marginalized suburban residents. The concept of a greater Parisian agglomeration challenged the traditional view of the city of Paris itself. On paper, it is fairly easy for a reader to visualize where the city ends. A thick line on a map confidently declares the limit of one jurisdiction and the beginning of another. On the ground, Thiers's wall had made the distinction quite clear for decades. Yet the continued influx of newcomers to the region rendered that once-formidable barrier less relevant each year. Bureaucratically, the Seine department administered Greater Paris. As *préfet* of the Seine during the Second Empire, Haussmann concentrated his efforts almost exclusively on Paris itself. He believed that the capital would eventually grow to occupy the entire Seine department and even surpass it in certain sections (Carmona 415). Under the Third Republic, however, lawmakers acknowledged that residents of the suburbs had a voice (and a vote) and, as in any functioning democracy, the government began to take into account their needs in the greater urban discussion.

To succeed, the growing capital needed to focus on regional-level reform. Urbanist Jean-Claude Nicholas Forestier felt that any 'réflexion sur la capitale ne peut plus se borner à la ville intramuros, car Paris et la banlieue constituent désormais 'une seule et indissoluble agglomération [...] "En un mot, le développement actuel de Paris, c'est le développement de sa banlieue, le développement de l'agglomération parisienne, de cet énorme groupement large de plus de 25 km"' (Cohen and Lortie 150). To be useful for the nation's capital, any plan to address the growing region had to encompass a much larger area than before. According to Forestier, any 'plan de la région de Paris doit atteindre un rayon de cinquante à quatre-vingts kilomètres, sur trois, quatre départements ou plus encore, sans tenir compte de leurs limites administrative. Il comprendrait: au nord, Chantilly; à l'ouest, Mantes, la forêt de Rambouillet; au sud, Dourdan, Melun; à l'est, Meaux.' In his 1906 *Grandes Villes et systèmes de parcs*, he viewed the Paris agglomeration as an 'énorme groupement large de plus de vingt-cinq kilomètres.' This perspective had changed considerably in the previous 25 years, when paying attention to 'Paris' meant focusing exclusively on the city itself. Forestier argued for the creation of a regional planning office comprised of technicians, urbanists, engineers, architects, lawyers, etc., to oversee various *études d'aménagement* under the purview of the Ministre de l'Intérieur (Gaudin 145). For Forestier, to create 'Le Grand Paris,' bureaucratic lawyers and planners, in addition to politicians, would have to come together to administer the metropolitan area.

To govern and manage the growing Paris region, French lawmakers had enacted various laws and regulations throughout the Belle Époque. In 1884, parliamentary legislation first acknowledged the need to decentralize the Paris region's administrative power, to take it away from the office of the *préfet* in order to foster more intercommunal cooperation. Camille Chautemps, a deputy from Indre-et-Loire, spoke of the increasing difficulty of regulating Paris, singling out the loi du 1884 as 'un monument législative d'une grand valeur à l'abri duquel [...] les communes françaises peuvent assurer la plénitude de leurs besoins matériels et travaux' (qtd in Gaudin 143–44). The actual wording of the decree gave the Conseils municipaux the ability to 'régler les affaires de [leur] commune' through mayors, though Paris obviously remained without one. The new power would 'exerce un droit de police, dont il rend compte non au conseil mais à l'autorité de tutelle. A côté des tâches de sûreté et de maintien de l'ordre sur la voie publique, cette compétence connaît un troisième visage, la réglementation de la salubrité' (qtd in Gaudin 144). After the chaotic nineteenth century, controlling streets and improving hygiene remained an important goal of city governance. Upon the turn of the century, the 1906 loi Beauquier created departmental committees to designate and preserve national landmarks. For what would become French urbanism, this latest piece of legislation established the right of towns to exercise eminent domain for reasons of public beauty (Horne 256). With this law, individual French communities gained more control and say over their own land and territory.

Fortifications – Potential Green Space

As officials debated the outer limits of Greater Paris, Thiers's fortifications remained a physical obstacle to any comprehensive plan. Following the post-First World War armistice, the French authorities returned their focus to urban conditions. Jean Claude Nicholas Forestier pessimistically described the state of Paris before the war:

> Depuis cinquante ans, malgré l'énorme accroissement de la population, aucun effort n'a été tenté pour lui donner de nouvelles facilités de circulation et de nouveaux espaces libres pour sa recréation et son repos. Le Paris moderne, dans son atmosphère de gaz et de benzine, se contente de l'oxygène qui suffisait à peine au Paris de l'Empire.

For Forestier, modernity represented not progress, but problems:

> Les usines, les voies ferrés, les gares de triage, entre autres, ont déjà étendu la main sur la plus grande part des terrains libérés et leurs exigences augmentent chaque jour. [...] Et pour couronner le tout, il trace un tableau consternant de la banlieue, 'immense ville concentrique en formation,' amoncellement d'épandages et de lotissements, [qui] établit aux portes de Paris la poubelle et l'encombrement en permanence. (Gaudin 141–42)

The polluted capital city seemed to choke under the weight of unsightly buildings and dirty industry. To compound this unpleasant situation, nearby suburbs boxed in the capital and limited growth. The lack of open space meant that there was no hope to bring in fresh air. Any repurposing of Thiers's wall, the only available avenue, represented a unique opportunity to reclaim new land, but also posed many administrative, jurisdictional, and practical problems.

The passage of the 1883 Guyot Proposition called for 'la désaffectation du mur d'enceinte, sa cession à la ville, dans des conditions à déterminer, ainsi que celle des terrains qui en dépendent, appartenant à l'État' (Conseil général 265). Thus, Paris had begun discussions for the transfer of control of Thiers's wall and its adjacent land from the army a little more than a decade into the Third Republic. For a city that could no longer expand physically, the potential new surface area was not insignificant. At 33 kilometres in circumference and 142 metres wide, this potential green belt offered an additional 250 metres of extra territory in front of the artillery installations where, in theory, it was forbidden to build. Once demilitarized, this territory would provide the city and region with a sizeable portion of undeveloped land in the middle of a large congested urban area. However, in reality the terrain was not actually vacant. The official 'no man's land,' the empty zone, cleared to provide line of site for the city's defences, had filled up quickly following the Franco-Prussian War. The area was a 'wasteland notorious for its ramshackle shantytowns inhabited by poor workers' and by 1926, the population of the zone was estimated at 42,000 persons (Horne 251). The barrier, host to slums with unsafe and unhealthy construction so near the capital, could not remain in its current state. Thiers's wall represented not protection but limitation and remained the last literal barrier to the capital's growth. The physical divide also became a mental one as it separated the *intra-muros* city from the territory outside. When Paris removed the Farmers-General Wall, additional boulevards lined with buildings appeared in that vacant space. The precedent, therefore, was to develop and build on the reclaimed land.

By the turn of the century, fortifications as a whole became obsolete for defence purposes in most European cities that still had them and governments across the continent began to remove city walls. Vienna and Cologne, two prominent examples, dismantled their inner fortifications and replaced them with *ringstrasse* developments: parks penetrated by buildings, promenades, and roadways. Between 1900 and 1912, the various stages of both cities' extension and modification plans were much publicized in France and were discussed and presented as desirable prototypes for Paris (Wolf 18–19). The Vienna case was especially intriguing. The capital and largest city of the Austro-Hungarian Empire had banned factories and limited new construction to only one-third of the former barrier's footprint. Furthermore, the Austrians incorporated the wooded hills around the city into the park system (Mabilleau 60). Forestier's description of Paris, above, painted a picture of a dirty city but Vienna's modifications proved that this need not be the case and even in a modern metropolis new parks could improve city life considerably.

In 1900, the French government first made public a plan to cede a western portion of the fortifications, adjacent to the Bois de Boulogne, to the city of Paris. Cries of protest erupted from sports enthusiasts, urban beautification advocates, historical preservationists, and hygienists interested in creating and protecting open spaces in the capital. The Parisian press, notably *Le Figaro*, denounced the 'amputation of the Bois de Boulogne' and Pierre de Coubertin, founder of the International Olympic committee, suggested forming a task force charged to preserve that segment of the woods (Horne 252). The harsh reaction to the idea of transferring just one section of the wall foreshadowed the larger debate to come for the entire wall.

As with any large-scale question with wide-reaching ramifications for millions of people, differences emerged in the wall debate. Government officials, epitomized by Paris municipal council member Louis Dausset, wanted to construct in the soon-to-be vacant land in order to expand Paris's tax base. They opposed members of the nascent French urbanist movement, many of whom championed the preservation of the highly desirable open land, as if it was economically unwise for the nation's budget. To hinder possible new construction there, preservationists recommended that potentially affected communities near the fortifications should acquire as much land as possible on their own in order to stop any unwanted building by private developers their jurisdiction. More importantly, once this space was in their possession, the preservationists encouraged town leaders to conserve, value, and not abandon it:

> Les municipalités doivent: 1. Garder jalousement en leur possession les terrains qui leur appartiennent, sans jamais les vendre. Ceux qui ne leur reviennent pas à un prix élevé et ne sont pas destinés à la création de parcs et jardins publics, doivent être loués à des baux de très longue durée et aux prix les plus réduits pour faciliter la construction d'habitations entourées d'espaces libres et permettant d'y loger à petits loyer la grande classe des travailleurs. 2. Les municipalités doivent saisir toutes les occasions pour augmenter leur domaine en achetant suivant les circonstances, même à de grandes distances dans les banlieues, des terrains à grande surface. (Gaudin 31)

Once local officials realized the benefits that open space near residential housing provided in reducing both pollution and disease, it would not be long before the national government did too.

Musée social and Eugène Hénard

The debate around Paris's fortifications centred on a desire to regulate the health and hygiene of Paris and other urban areas and grew in part from the budding influence of the Musée social. On 19 May 1894, a group of prominent preservationists met at the home of the Comte de Chambrun and officially established the Musée social. Founded originally as a forum for research based around documents from the 1889 Exposition Universelle's social economy section, one of ten official groups that constituted the large-scale event, it became the focal point of studies of hygiene, social reform and, ultimately, urbanism (Charvet 225). The organization established three tasks for itself: to reinforce the networks of associational life; to inspire an enlightened elite to actively pursue solutions to the social question; and to assemble a broad-based documentation on social movements in France and abroad (Horne 91). In addition to industrial welfare and labour management, the 1889 social economy section also set out to merge the social and political heritage of the French Revolution.

As it matured, the Musée social focused its energies towards providing better housing and enhanced access to clean water and fresh air for all socioeconomic strata. In terms of French urban development, the group also became politically active after the turn of the century. The group improved the lives of millions as it influenced French urban reform in the first half of the twentieth century (Wolf 17). The fortifications debate galvanized members to act and epitomized the primary struggle faced by the Musée social. The construction of new parks in the liberated territory would undoubtedly improve the lives of residents nearby, but business leaders had considerable sway in French politics and would benefit significantly from new buildings in the affected zone. The Musée social, armed with architects, joined forces with like-minded politicians to fight development in the fortifications and became anchors of a budding anti-building, pro-green space coalition.

In 1905, the Musée social entered the political arena. It formed an unofficial partnership with politician Charles Beauquier, a radical-socialiste who championed environmental causes. Beauquier's work in the Chamber of Deputies helped the group prepare for their eventual role in upcoming municipal elections three years later. This 1905 collaboration also coincided with an International Congress on Tuberculosis held in October of that year. The large consortium of scientists gathered in Paris to fight disease further highlighted the Musée social's message of improving health and hygiene for the poor. The Musée social also engaged the media and editors from newspapers, such as *Le Figaro* and *Le Matin*, who joined the campaign against substandard development on behalf of the hygiene movement (Charvet 230). This political and media support helped the Musée social broaden its reach and expand its work into new ventures.

On 6 January 1908, the group's director, Léopold Mabilleau, announced that the Musée social's board of directors had created a Section d'hygiène urbaine et rurale whose goal was to work on 'tout ce qui intéresse directement la vie familiale des travailleurs (habitations, jardins ouvriers, espaces libres, alimentation, etc.)' (56). The Musée social's hygiene section initially concentrated on smaller projects, such as replacing the decommissioned Saint-Lazare prison, an endeavour that epitomized their mission. Jules Siegfried, the Musée social's president, was troubled that, in the neighbourhood surrounding the prison in the tenth *arrondissement*, 'il manque absolument, non seulement de parcs, mais même de squares, et qu'il faut s'efforcer de décider le Conseil municipal à ne pas aliéner les terrains qui vont être rendus disponibles pour un autre usage que celui de jardins' (Mabilleau 57). The Section d'hygiène also became involved in the partial demolition of the old Temple market in the third *arrondissement*. However, instead of creating a small park there as the Musée social preferred, the city sold the land to developers who then constructed a six-storey apartment building. This result further impressed upon the Musée social's members the need to change how the French addressed open space in the city. An additional dedicated call to action centred on the subdivision of the land adjacent to the Champ-de-Mars, which eliminated a quarter of the open space near the Eiffel Tower. The Musée social leadership feared that such a pro-business action would set a dangerous precedent and, if the same thing happened to the land containing the fortifications, it would mark a defeat for the group's mission and for the residents of the area who hoped to benefit from the use of that land.

Prior to the First World War, the Musée social drew up and publicized specific plans for urban improvements, such as their work on the political posters and recruited experts in the field of architecture. Alfred Agache, Henri Prost, Robert de Souza, and Léon Jaussely had all developed national and international reputations for their work in urban design and contributed to the Musée social's body of work. Louis Bonnier, Inspecteur général des services techniques d'architecture et d'esthétique de la Préfecture de la Seine, proved to be a useful ally to the cause. Because of his position within the city's administration, Bonnier had direct ties to the Seine's *préfet* and served as a conduit for the flow of ideas that developed within the Section d'hygiène into Parisian government planning circles. As the Musée social grew and became more active each year, increased press and publicity also further strengthened its name recognition and brought in new members. Among them was the architect Eugène Hénard, an employee of the Paris municipal Bureau of Public Works.

Beyond Jaussely, it is perhaps Hénard, one of the most influential French urban architects of the time, who provided the most detail and substance to the Musée social's vision. For example, the Section d'hygiène chose his plan for its 1908 campaign to transform Paris's fortifications into parks and Hénard's proposal appeared on a series of 12,000 posters placed in the Paris metro and throughout the city (Horne 256). Furthermore, his expertise in urban planning bolstered the Musée social's credibility. Hénard earned international stature as an urbanist for his series of studies, published between 1903 and 1908, on the renovation and subsequent extension of the city of Paris. Hénard's Musée social plan for the fortifications featured the creation of nine large public parks plus 13 playing fields along the urban periphery as well as a boulevard – *la grande ceinture* – to encircle the city. Hénard suggested repeatedly that Paris needed to prepare a plan for its *extension et aménagement* which, in turn, should then be coordinated with a plan for the development of the region surrounding the city as well. Acceptance of the idea that Paris should have a master plan to guide and coordinate its development became one of the principal objectives of the Musée social. In 1907, at the first meeting of the Section d'hygiène, members agreed that cities needed to plan on a more regional level. In many communities, Siegfried lamented, 'c'est à peine s'il existe un règlement de voirie' (Mabilleau 57). Hénard's ideas, the Musée social argued, would work on a practical level and represented the public good since they promoted not just traffic, but also public hygiene. They asserted that any proposal that simply handed the fortified zone over to private real estate speculators should be rejected. Jules Siegfried and Ambroise Rendu, vice president of the Conseil municipal de Paris, promoted Hénard's plan and called on the state to directly concede the land that held the fortifications to the city, sacrificing any profit it might make since, as they argued, 'it is in the national interest to make Paris the most beautiful city in the world' (Horne 256). Beauty and aesthetics, therefore, should trump commercial development. For urbanists around the time of the First World War, Hénard represented how hygienist ideals could work in the real world and influence urban design.

Between Haussmann's large-scale transformations in the nineteenth century and Le Corbusier's unrealistic ideas in the 1920s, only Eugène Hénard presented plausible, if not still radical, proposals for the next step of the capital's modernization: a combination of social objectives and urbanistic thought that would come to be known as 'the French school of urbanism' (Horne 254). Hénard had little actual building experience, but on paper he was rather prolific and presented his vision in various publications. He published two influential pamphlets: one on the overall greater fortification question, the other on open urban spaces. He published two influential articles in *L'Architecture* – 'Les Espaces libres et les fortifications' (1908) on the overall greater fortification question, and 'L'Embellissement des villes' (1910) on open urban spaces – which solidified the new ideals that would help form the basis of French urbanism in the first half of the twentieth century. In 'Les Espaces libres,' Hénard responded to architect and builder Paul Lafollye's critique of both the Paris municipal council member Louis Dausset, who also submitted a plan for the fortification zone, and the Musée social. In his defence, Hénard championed anew the Musée social's goal of preserving more available green space and distinguished it from Louis Dausset's proposed use of the arc of territory, which called for 'un lotissement rationnel' (Dausset *Résumé*, 338). According to Hénard, in Dausset's plan, the green space adjacent to the wall would in reality be 'coupée et recoupée dans tous les sens par les voies de sorties de Paris, et sa surface, quoique plus grande, sera moins bien protégée que celle des neufs parcs dont les emplacements ont été soigneusement choisis en dehors du passage des routes fréquentées' ('Les Espaces' 336). More specifically, Hénard trumpeted the Musée social's creation of almost 168 hectares of parks, gardens, and open space which, if built, would increase the amount of green space in the city by almost 60 per cent. Further aligning himself with the Musée social's goals, Hénard cited the hygienic effects of such new land:

> C'est un morceau de terrain aménagé, comme un morceau de campagne, plus aéré, plus ensoleillé qu'une place publique, dans

> lequel la végétation épure et bonifie l'aire qu'on y respire, où les vents et les brises renouvellent cet air plus rapidement et plus complètement que dans une enceinte entourée de bâtisses, où les massifs de feuillages arrêtent et filtrent les poussières malsaines et les germes contagieux. ('Les Espaces' 335)

Not only would less construction allow for the planting of new vegetation, but wind would filter the air and rid it of germs and dust as it moved through plants and trees.

Economically, Hénard believed that even undeveloped land had significant value and the city should not just give it away to the highest bidder. He was afraid that the potential sums provided to the government coffers would grant extra favour to businessmen and developers, and in turn, ignore inexpensive or unregulated building in open spaces as had happened in the past. In his response to the developer Lafollye, Hénard questioned the idea that 'la suppression de la servitude militaire libérerait 300 hectares de terrains zoniers, sur lesquels on pourrait édifier 7500 immeubles d'une surface moyenne de 400 mètres chacun, c'est-à-dire presque un dixième en plus des immeubles actuels de la capitale.' Hénard felt that buying the land cheaply, developing it, and then selling it for a profit would benefit too few. Instead of constructing more towers to reduce rents for the working poor, the city should instead pass laws to regulate it. Lafollye had lectured Hénard, and perhaps the Musée social by proxy, on the benefits of such an arrangement: 'M. Hénard a-t-il réfléchi à la répercussion énorme de ce résultat sur la richesse immobilière actuelle des propriétés de la banlieue, et à la concurrence désastreuse qui viendrait frapper les terrains de l'enceinte acquis par la Ville de Paris et lotis à grand frais' (Hénard, 'Les Espaces' 337). The business community had a lot of influence, but not all developers were convinced that Paris could build its way to perfection. In an essay for the *Journal de la Chambre des Propriétaires* published on 16 September 1909, Léopold Mourges the secretary of the Ligue des Espaces libres and also of the Chambre des Propriétaires – hardly a militant leftist – voiced support for Hénard's philosophy. He postulated that creating parks would benefit the capital just as much as new apartment buildings. In his commentary, this secretary of a group dedicated to new construction proposes developing the fortification land, but for recreational use:

> Ne vaudrait-il pas mieux voir grand – nos devanciers n'ont jamais vu trop grand et trop beau pour Paris – et conserver en espaces libres les fortifications et la zone pour y établir des parcs, terrains de jeux, etc. qu'on agrémenterait, selon les besoins, de monuments d'utilité publique? Ce serait la suppression de la spéculation que plus grand profit de la généralité des Parisiens, *qui n'aurait pas à débourser plus de 300 millions* [...] pour accomplir cette grande œuvre. (Hénard, 'Les Espaces' 337–38)

Although multiple new towers would be good for certain businessmen, Mourges believed, like Hénard, that Paris and Parisians would benefit more from improved access to green space.

Others did not feel that the Musée social's philosophy, in its current form, was what Paris needed. Louis Dausset had his own ideas.[29] His project, supported by Hénard, differed from Siegfried's Musée social proposal in that it focused on the entire ring of fortifications, not just the portion affected by the 1898 law. He expressed his viewpoint on the fortification debate to expose the futility of the critics' ideas as well as the 'impossibilité d'aboutir à un résultat plus favorable tout en tenant compte des moyens financiers de la Ville. En outré il précisa l'économie de son projet et les conséquences heureuses qu'il était permis d'en escompter' (*Résumé* 337). Dausset, on the other hand, felt that Siegfried's proposal was too complex and therefore a non-starter as it would only slow down negotiations. Even though Siegfried's ideas came from his position as a deputy, for him, Paris needed a large comprehensive solution as the wall represented the 'dernière réserve d'air naturelle qui s'offre à nous' (Charvet 252). Dausset, too, eventually fell in line with the Musée social's Section d'hygiène and incorporated many of its ideas into his own plan (Charvet 283). Dausset saw the dismantling of the former barrier as an 'importante affaire qui intéresse à un [...] haut point de l'hygiène et la beauté de Paris' (Charvet 249). When Dausset submitted his own plan, the Ministry of Finance's Commission chargée d'étudier les questions relatives au déclassement des fortifications de Paris opened up three separate sub-commissions

29 On 4 November 1908 Louis Dausset presented his own *projet d'aménagement pour les fortifications*. The city would buy the fortifications outright, they would be subdivided, and then transformed into open space. Any leases would be honoured until they ran out and the city would annex all the remaining land (Charvet 244).

to study the terrain, acquisition possibilities, and associated financial consequences. Each one spent time investigating its charge, wrote reports and, by December 1912, they had all presented their findings to the original Commission (Dausset, *Résumé* 338). In addition to the fortifications question, the city needed to consider how to manage current vacant lands beyond the zone, which led back to the original pre-1915 task of ordering growth in Cornudet's proposal. In further agreement with Siegfried and Hénard, Dausset contended that, when deciding the future of Paris, lawmakers must look beyond the city limits. His plan constituted 'une opération de salubrité publique, pour le bien de trois millions sept cent mille habitants qui, avec les communes de banlieue, forment l'agglomération parisienne' (qtd in Charvet 253). He sympathized with the needs of the suburban communities. With limited room, they had no choice but to build where they could find available land:

> Depuis cinquante ans, l'accroissement de leur population a nécessité la construction de nombreux immeubles, et par le fait même la disparition de leurs réserves d'espaces libres, si bien que très peu sont encore dotées de promenades ou de parcs publics d'une étendue susceptible de suffire aux besoins de leurs habitants. (Dausset, *Résumé* 339)

Although suburban residents remained a factor in his plans for the region, Dausset's goal was to improve Paris and the agglomeration as a whole, not just certain towns and villages.

Hénard's ideas were prominent in Paris's urban discourse, and his positive views on automobile movement influenced future planners. Several young Frenchmen with experience in urbanistic planning served under him. Most important among them were Alfred Agache (1875–1915) and Henri Prost (1874–1959). On behalf of the Musée social and with the help of Agache and Prost, Hénard prepared one of the first comprehensive plans for the improvement and extension of the city of Paris as the centre of the greater region. Beyond the access to parks and open space, Hénard's *Études sur la transformation de Paris* (1904) brought the subject of roads and transportation to the Paris urban debate. His plan included many of the suggestions he had made between 1903 and 1906, such as neighbourhood parks and a *boulevard à redans* along the fortifications, new transversals for the city, and new radial roads penetrating from the *noyau central* to the surrounding countryside. But it also indicated large areas of reserve land for future parks throughout the region if Paris was to expand from 7,883 to 19,500 hectares, and at the northern and southern edges of the city there were large airfields (Wolf 84–85). So, to manage Paris's growth, Hénard agreed that one needed to look at the larger region.

Local Urbanism

The First World War temporarily deferred any regional planning questions. However, once hostilities ended, the fortifications debate returned and the government discussed what it should do for French communities, especially those which had suffered war damage. The loi Cornudet, first introduced in 1915, became law on 14 March 1919, just months after the November 1918 armistice. Debate, prolonged by Musée social, helped direct this new legislation. Regulations within the loi Cornudet created France's École des hautes études urbaines. That institution's founders, among them Marcel Poëte and Henri Sellier, specified the mission to teach this new science: 'Le Conseil général de la Seine, par une délibération du 9 juillet 1919, prise sur la proposition de M. Henri Sellier a créé un enseignement public des questions urbaines près de l'Institut d'histoire, de Géographie et d'Economie Urbaines de la Ville de Paris' (Van Damme 232). Thus, as politicians passed France's first urban law, they also formed an institution to train French students in the new discipline of urbanism. These urbanists would set the tone for French urban expansion from this moment forward. As French cities created and published their *plans d'extension* in the early twentieth century, various terms used in conjunction with 'urban' appeared with more regularity in the press and in official documents: 'voirie, travaux publics, classement des activités, normes d'hygiène, permissions de construire' (Gaudin 148).

In the eyes of public officials, the 14 March 1919 law gave authority to French urban areas of a certain size. For a nation used to dictates and regulations coming from Paris, this transfer of power marked a significant change. Now urbanists, essentially non-politicians, could apply their new discipline to cities and help influence the planning process, a trend that already existed in England, the United States, and Germany. In theory, science, not politics, would now transform living conditions for rich and poor. But there were problems with the law. Henri Sellier felt that it was necessary to address the government's limited ability to compel the application of the new regulations by unwilling towns:

> Si nous voulons que la loi sur les plans d'aménagement et d'extension ne reste point, comme tant d'autres, sans application et sans portée pratique, mais qu'au contraire, elle pénètre l'action administrative et fasse bénéficier la population des grands centre d'une amélioration permanente de leurs conditions d'existence, il importe que les représentants de l'agglomération parisienne [...] prennent l'initiative de coordonner les bonnes volontés et les efforts dispersés, d'établir l'unité indispensable de vues, de susciter un mouvement d'études et de réalisations et d'assurer enfin la discussion de celles-ci. (Van Damme 233)

Any discussion of a regional plan needed to involve people of all levels and have them work together for the good of both the city and suburbs.

Paris's Concours pour le plan d'aménagement et d'extension de Paris was one of many throughout France but since it concerned the capital, it received the most attention from both national leaders and the press. Other *communes* also looked to revamp their suburban growth and modify existing infrastructure. Seven years before the loi Cornudet came into effect, Lyon became the first city to undertake large-scale urban renovation. Although not as complex as Paris's needs, the sizeable city on the Rhône and Saône rivers called for the creation of a radio-centric formation in which angular streets were to be superimposed on Jean-Antoine Morand's eighteenth-century plan. In Lille, the debate over the declassification of the city's fortifications coincided with Paris's and led the city to open up its 'projet d'aménagement et d'extension' to competition as well. In 1920, Jacques Greber won the right to design the *préfecture* of the Nord department's *plan d'extension de la ville*. In his proposal, he projected a ring of HBM, gardens and green spaces in the periphery (Delorme 14). In Besançon, native son Maurice Boutterin was charged with creating a *plan d'extension* that incorporated the city's uneven terrain, which included sweeping views, as well as Vauban's fortifications (Delorme 17). Marseille, on the contrary, remained an urban renewal outlier in many respects. In 1906, it organized one of France's first urban competitions for its *quartier* of la Bourse (Delorme 18). Yet no region-wide plan was approved until the 1931 Plan d'extension et d'aménagement de la ville de Marseille, making it one of the last large French cities to fulfil the 1919 loi Cornudet's requirements.

The government had specific ideas of what it wanted to see in the capital and its environs. Louis Bonnier, Inspecteur général des services techniques d'architecture et d'esthétique de la Préfecture de la Seine, authored an article entitled 'L'Esthétique et l'architecture dans le concours pour le Plan d'aménagement et d'extension de Paris,' for the 1920 issue of the journal *La Vie urbaine*. In that text, he explained that the principals of urbanism must be present in any plan to transform the Paris cityscape. Ideally, the new Greater Paris would be able to combine the best of both aesthetics and architecture:

> Si on ne comprend pas l'esthétique au sens restreint qu'on lui donne d'ordinaire, [...] on peut concevoir qu'une cité où les rues seraient larges, les maisons confortables, les monuments pratiques, les promenades importantes, se rapprocherait de l'idéal de nos amis outre-Atlantique qui veulent leurs villes commodes pour le travail, économiques pour les affaires, saines pour la vie de famille, agréables pour la vie sociale. (Bonnier 29–30)

Architecture, an equally important component, should further add to the city's draw: '[il] ne joue qu'une partie intéressante, mais relative, dans l'immense orchestre de l'urbanisation d'une ville' (Bonnier 30). The incorporation of the past architectural grandeur of Paris, of which Haussmann's legacy loomed large, needed to be reconciled with what the city required in order to move forward. Opening up Paris's plan to competition would demonstrate how multiple urban experts envisioned cities of the future: 'Les lauréates des quatre sections sont, à de rares exceptions près, surtout en ce qui concerne les sections d'aménagement et d'extension, des architectes, mais ce sont essentiellement des urbanistes ayant ajouté à leur prédis positions professionnelles les connaissances multiples que nécessité la science des villes' (Bonnier 31). In addition, competition would provide free publicity for Paris, showcase French ideas, and demonstrate in which direction French urbanism was headed.

On 10 April 1919, the Conseil municipal de Paris officially invited the Seine prefectorial administration to present the requirements for Paris's Concours by 12 July of that year. They determined that each entrant would need to include two items. The first would serve as the main draw, a 'wish list' of projects and improvements to the capital. It called for a map

> qui fixe la direction, la largeur et le caractère des voies à créer ou à modifier, détermine les emplacements, l'étendue et les dispositions des places, squares, jardins publics, terrains de jeux, parcs, espaces libres divers, et indique les réserves boisées ou non à constituer, ainsi que les emplacements destinés à des monuments, édifices et services publics.

The second component would present the 'bones' of the details needed to complete any potential regional design. In the plan, the Conseil needed to see a

> programme déterminant les servitudes hygiéniques, archéologiques et esthétiques, ainsi que toute les autres conditions y relatives et, en particulier, les espaces libres à réserver, la hauteur des constructions, ainsi que les prévisions concernant la distribution d'eau potable, le réseau d'égouts, l'évacuation et la destination finales des matières usées et, s'il y a lieu, l'assainissement du sol. ('Le Programme et le résultat du concours' 91)

The contest opened on 1 August 1919 and, to provide each entrant with a level playing field, the city of Paris furnished a copy of prior legislation that affected the city along with nine maps which they could use for their presentations.[30] Access was granted to the Bibliothèque administrative de la Ville de Paris as well as the Bibliothèque de l'Institut d'Histoire, de Géographie et d'Economie urbaines to help with research ('Le Programme et le résultat du concours' 99). This information provided the substance for most plans and helped ensure that entrants employed correct, up-to-date information.

The jury, made up of a variety of politicians (including the *préfet* of the Seine, three senators, five deputies from the Assemblée nationale, six Paris municipal council members, the *préfet* of the Police, etc.), bureaucrats, and architects ensured that each competitor adhered to the wording of the 1919 law but also that they

30 Each entrant received a copy of the following maps: 'Un plan de la région parisienne, en couleur, au 1/50.000e'; 'Deux plans de la région parisienne, en noir, au 1/40.000e'; 'Deux plans du département de la Seine au 1/20.000e'; 'Deux plans de Paris au 1/10.000e'; 'Un plan de Paris et de ses alentours au 1/20.000e'; 'Un plan des fortifications et de la zone au 1/5.000e.' In addition, each candidate would receive three sheets of their choice from 'le plan de Paris au 1/5.000e' ('Le Programme et le résultat du concours' 99).

> entreprendre l'étude demandée avec les vues les plus larges, sans se préoccuper des circonscriptions administratives, et en englobant dans leur plan des communes du département de la Seine, et même, s'ils le jugent nécessaire, des parties des départements voisins en un mot, tout ce qui, dans leur conception, constitue l'agglomération parisienne, non seulement telle qu'elle peut être définie à l'heure actuelle, mais telle qu'ils en imaginent le développement et l'extension dans l'avenir. ('Le Programme et le résultat du concours' 92)

Each entrant could decide where the edge of Paris lay. This rule provided them with the flexibility necessary to complete a plan for the region's future.

The government created four contest sections. The first, Section I, was the most important as it dealt specifically with requirements of the 14 March 1919 law. The entrants for this competition concentrated not just on Paris, but on the 'l'agglomération parisienne tout entière.' Section II was reserved for projects that embellished the interior of Paris alone. Section III was reserved for those who had projects regarding the fortifications only. Section IV included any entrant who wished to address only a part of the city or agglomeration ('Le Programme et le résultat du concours' 97). In all, 60 entrants submitted plans to the panel of judges.

Although not present in the current competition, Hénard loomed large. Bonnier, an advisor to the city, believed that the most important component to any successful metropolitan area, especially for Paris, should be based on Hénard's vision for the city: 'c'est le moyen de transport en commun propre, rapide, fréquent et à grande portée, traversant la banlieue actuelle, qui est encore la ville, pour gagner les bois, les forêts, les espaces lointains' (Bonnier 31). As Hénard had previously recommended, a successful city needed to enhance the movement of people and goods. He described how 'dans l'aménagement intérieur de Paris, les uns ont prévu quantité et percements nouveaux, de grandes traverses larges et coûteuses du sud or nord, de l'est à l'ouest, avec des avenues de 75 et même 150 mètres de largeur, analogues à l'avenue du Bois de Boulogne ou au cours de Vincennes' (33). Large 'piercings' *à la* Haussmann remained the go-to method of freeing the city from dark, crowded streets. Yet, since it focused on expensive, oversized prestige avenues, this typical Parisian version of an urban plan unsettled current officials. Bonnier lamented the fact that 'la méthode française préfère attendre que les besoins se manifestent, ou même se soient produits depuis longtemps, pour entreprendre de les satisfaire et ne cède le plus souvent que devant des réclamations réitérées' (Bonnier 33). Instead of looking to the past for answers, Bonnier recommended instead anticipating what the city might need in order to grow in the future. For example, the desire to simplify the cityscape remained the top priority. Superfluous additions, such as skyscrapers, new monuments, and street furniture could and should be avoided, in preference for improved *entrées*, traffic circles, and promenades. The latter categories would facilitate movement and therefore improve commerce and growth.

The Commission d'extension held Paris's exhibition at the Jardin d'Acclimation and welcomed architects and urbanists to view the exhibitors. Bonnier summarized many of the main themes present in the 60 plans submitted. Overall, as the Commission wished, entrants worked to reduce density and improve access to nature and open spaces. Most projects focused on transportation. Some contestants wanted to empty central Paris of housing, similar to the City of London, the financial district that is bustling and lively during the workday, but empty at night and weekends. Beyond density, Bonnier pointed out two principal differences among the entries. One group proposed creating many new streets and 'larges et coûteuses' avenues, while the other half proposed using Paris's existing infrastructure, just improving and widening them (Bonnier 33). Choosing the latter may have helped their projects in the eyes of the judges as post-war rebuilding meant a smaller budget for new construction in the short term.

Architect Léon Jaussely, an entrant, had his own specific ideas on urban reform. In his editorial entitled 'Chroniques de l'urbanisme' for *La Vie urbaine*, a journal he helped found, he espoused the tenants of the new field of urbanism and how he thought it could improve the city and surroundings. The goal of the journal, created by the Conseil Municipal de Paris, was to provide a comprehensive look at 'des phénomènes qui président à la formation originelle des agglomérations, puis à leur développement' (181). Jaussely, whose plan would ultimately be chosen by Commission, asked a very important question:

> Quelle est la limite de ce que l'on peut appeler l'agglomération parisienne. Jusqu'à ce jour elle est encore indéterminée, aucun

> tracé n'en a été fait qui soit appuyé sur des résultats fournis par la science. Entre l'extrême limite de la région économique sur laquelle, corps de pieuvre, Paris étend ses tentacules, d'un coté en aspirant les ressources, mais de l'autre lui apportant la richesse, et le Paris enserré en ses fortifications, il existe une zone moyenne qui vit et procède complètement de la capitale qui lui donne et lui doit tout [...]. Pour une organisation sûre de ce qui est Paris aujourd'hui, il est indispensable de connaître ses limites; les études entreprises permettront d'y parvenir. (183)

As the Parisian agglomeration grew, it changed shape and needed to be understood and controlled. But since science could not be used to determine Le Grand Paris's limits, one had to find another way. Until Paris officials defined that boundary, how could one manage the limits of a growing agglomeration? Jaussely warned that,

> Pour un plan d'extension, il faut, non seulement connaître ces limites, mais encore partir de là pour se hasarder à en tracer d'autres – comme l'a fait M. Bonnier dans son article de cette Revue sur le Grand Paris en 1861 – qui les dépassent parfois largement; ces hypothèses toutefois doivent être aussi peu hasardées que possible, car les œuvres de travaux publics: routes, boulevards, promenades, parcs publics, chemins de fer, métropolitain, tramway, etc. sont couteuses à établir et à entretenir. (184)

A regional definition by population, considered arbitrary by some experts, stems from a need to establish some sort of limit. The legislation required communities with more than 10,000 residents to create a comprehensive plan for development. It was hoped that studying the new science of urbanism could create orderly growth in a somewhat chaotic new urban order. Jaussely believed that the passage of the law could help maintain open spaces and also introduce the possibility of construction regulations by zone. Perhaps in homage to the Musée social, for Paris, this definition meant that urbanists could now target the development of the large apartment blocks that ruined the harmonious nature of the cityscape – avenues and monuments – and maintain the historical tone of certain neighbourhoods, streets, and squares. Jaussely expressed a desire to change the perception of an administration that tended to be more interested in preserving the past than in planning for the future:

> Nous ne trouverons rien à dire à cet article car il donnera plus de poids à l'administration municipale pour la conservation des sites urbains que n'en donne à la Ville de Paris le loi du 13 juillet 1912 complétant l'article 4 du décret du 26 mars 1852, et qu'il permettra les développements rationnels de l'urbanisme moderne. (195–96)[31]

The 1912 ordinance on the removal of the fortifications, which allowed for the funding of large-scale HBM construction, had diminished the aesthetics established under Haussmann and the Second Empire. So any future reconfiguration of the zone would need to take into account how new building there would fit in the overall cityscape.

Circulation

Jaussely's *plan d'extension* addressed Louis Bonnier's main points 'circulation, hygiene [et] l'esthétique,' the three principal conditions necessary for a workable municipal plan (29). How did Jaussely present these requirements to the Commission and to the French public and include what, he felt, Paris needed? All three aspirations appear on his map via new roads, parks, and open spaces as well as improved living conditions for suburban residents. When looking at the centre of his map, the first thing one notices is the city of Paris itself. The crowded capital blends in with the expanded surrounding suburban *communes*, divided only by the thin green line representing the area adjacent to Thiers's fortifications. Jaussely chose to present the urban fabric in a colour that did not differ significantly from undeveloped land.

The Musée social movement's involvement in the fortification debate exemplified its role in improving living standards for the working class. For urbanism as a whole, that amelioration meant access to parks, green space, and fresh air. However, one additional problem plaguing the city was transportation. As Paris street traffic increased, congestion within the city grew worse every year. The automobile, although faster, did not eliminate pedestrian or horse-drawn traffic overnight. It simply compounded the number of vehicles on already

31 Jaussely referenced article 1, which describes the character of the urbanization legislation.

Figure 6. Jaussely, Expert, and Sellier, 'Projet lauréat de la section générale du concourse du plan d'extension de Paris' (Plan d'extension), 1919.

crowded streets. Hénard had championed improving the movement of people and goods in his *Études* (1904). Technology also changed how people interacted with the city. Many urbanists and bureaucrats believed that the first underground *métropolitain* line, inaugurated for the 1900 World's Fair, represented 'la vraie solution de la circulation et de l'hygiène de Paris,' but Hénard himself was not convinced as the *métro* carried only pedestrians, not merchandise (Olson, 'A New Paris' 433). In many respects, the subway line did present a modern solution for moving people throughout Paris, even if it was not a brand-new concept in European cities. London started operating underground trains in January 1863. Improving transportation in the crowded city had been a high priority for French leaders since as far back as Philippe Auguste's twelfth-century goal to pave Parisian streets. Haussmann's boulevards represented the most recent solution to unclog the city centre and the Second Empire *préfet* had focused many of his transformations on moving people through the crowded medieval core. He also improved access from the newly constructed main train stations to the embellished centre of Paris. However, the *métro*'s reach only went so far as its routes were fixed and could not be extended or modified without considerable construction and cost. When business and industry needed to deliver goods to the outer *arrondissements* or the suburbs, the underground rail system provided no assistance.

The Commission believed that an enlarged road network should be one of its priorities. Jaussely included new roads, which also underscored his former mentor Eugène Hénard's desire for improved transportation. In the nineteenth century, Haussmann had demonstrated how wide new streets could replace insalubrious property plots and facilitate modern circulation patterns, and both Jaussely and Hénard followed this philosophy. Once Paris's streets had been improved, Commission members sought planners who would branch new roads and highways throughout the Seine department. Transportation-wise, Commission members felt that the train network centred on Paris, thanks to an 1841 law, should also be targeted. The Petite Ceinture railway, which ran parallel to the soon-to-be-defunct fortifications, needed to be removed in order to make way for more development, while the Grande Ceinture, a second rail belt built 15 kilometres outside of Paris to help alleviate rail traffic into the city, might be able to replace it.

As residents and businesses followed growth into the outer suburbs, transportation remained a key issue and the Commission looked for better access between the capital and the rest of the Seine department. Bonnier himself pointed out how 'le rond-point de la Défense, la Porte Maillot, le Mont-Valérien, [et] le cours de Vincennes' all served as focal points in urban improvement as they had helped ameliorate access to Paris. To the west of the city, via a green ceremonial *entrée* to the city, Jaussely acknowledged the importance of these avenues and proposed to lengthen the Champs-Élysées out to the La Défense via the Porte Maillot. This sector, named after the La Défense de Paris statue that overlooked the city, would anchor urban renewal on this edge of the region. The wide, green thoroughfare jets west from a crowded capital and facilitates access to the western *arrondissements* as well as allowing for a single, straight entrance from western France. The La Défense area, home to Jaussely's proposed new Exposition Universelle site – a move from its traditional home on the Champ-de-Mars – stands out on the map, its darker tint indicating its importance to the reader. To enhance access, Jaussely encircled it with wide routes. Visitors to the World's Fair site, whose cars created Hénard's 'la circulation exceptionnelle,' would now enjoy much easier access.[32]

A second principal route nearby takes potential drivers south-east, towards large green woods similar in size to the nearby Bois de Boulogne. Other embellished throughways branch out from Paris as well. Jaussely provided the eastern half of the Seine department with seven roadways leaving the capital as well as multiple ring roads that would allow drivers to avoid the crowded city centre if travelling around the metropolitan area without an excursion into the city itself. Unfortunately, with post-war repairs underway in northern France, the budget remained a concern. Bonnier preferred planners to concentrate on improving existing roads rather than build new ones. He advised urbanists to be aware of

> [les] dépenses et se rappeler qu'une voie nouvelle, souhaitable mais que son coût empêche de réaliser jamais, est moins intéressante qu'une rue améliorée; que à cause des pentes, le meilleur chemin

32 Hénard outlined six types of traffic patterns in the city. The sixth and final one, 'la circulation populaire ou exceptionnelle,' materialized around large-scale events such as exhibitions or visits from foreign heads of state (*Etudes* 182–83).

> d'un point à un autre point n'est pas toujours la ligne droite; que l'alignement rectiligne gâte souvent l'aspect d'une voie ancienne sans profit appréciable. (37)

Many intellectuals, on the other hand, privileged new construction. Foncin noted that a city grew by following 'les voies de communications. [...] Les faubourgs se sont d'abord bâtis en bordure des routes rayonnantes qui partent de Paris. Aujourd'hui les centres nouveaux jalonnent les voies de chemins de fer. Là où les rapports avec le cœur de la ville deviendront faciles et rapides s'élèveront des quartiers neufs' (82). Therefore, any new rail line or road could foster new construction away from the busy centre. If residents were given the opportunity to quit squalid crowded neighbourhoods for more space and could do so easily, she believed they would. Plus, those residents had a better chance of living a healthier, happier life. A few of Jaussely's roads on the outskirts of the agglomeration even follow train tracks since he knew that people would migrate in those directions. One such example can be found past Versailles and a second example follows a canal and rail line to the north-east of the capital. The rest remain independent of France's rail network. In an obvious nod to his preference for automobile traffic, Jaussely kept Paris's streets free of competition for drivers. He proposed to construct new rail lines within the city, but placed them all below ground. A central station under Les Halles would anchor the new *intra-muros* network. Dotted red lines, earmarked in the legislation as a 'partie à créer,' mark these underground rail links between existing train stations, with interconnections near the new *gares* at the Bourse, the Archives, and the Place de la République, which are also among the few markings inside the city.

If cars symbolized the future of individual travel to many planners, such as Jaussely, then public transportation, trains, subways, and tramways, on the other hand, represented the present mode of transport for the vast majority of Paris residents. The ever-expanding reach of suburban development meant that it took longer to travel the same distance each year. In his 'Chroniques de l'urbanisme,' Jaussely addressed the increasing amount of time needed to cross the city and region:

> Une nouvelle étude, que nous pouvons annoncer, a été entreprise par M. Bonnier pour la *Section de Géographie* de notre Institut d'Histoire, de Géographie et d'Economie urbaines. Elle a trait

> encore au développement de Paris dans le cours du XIXe siècle, à celui de la périphérie et de sa banlieue (grande et petite) en regard du chemin parcouru eu une heure au plus, par les voyageurs, selon les moyens de transport empruntés. (182)

Jaussely noted that as Paris expanded along those communication corridors and beyond, the suburban communities themselves also grew. This additional enlargement only added to the length of daily commutes as more people had to use and share trains, trams, and roads.[33] In order to widen the geographical area in which one could travel within a one-hour time period, planners would need to include more infrastructure. For a start, additional direct rail lines offering more frequent service would help people circulate more easily. Any such improvement would better unite the region's edges and improve the lives of those workers who had to commute every day, often changing trains multiple times to do so. Foncin echoed Jaussely's vision. She stated that 'il suffit, comme le projet Urbs l'indique ingénieusement, de doubler les grandes lignes existantes par des lignes nouvelles qui traverseraient les communes à urbaniser' (83). Bonnier continued this thought: 'Le seul et capital facteur, c'est le moyen de transport en commun propre, rapide, fréquent et à grande portée, traversant la banlieue actuelle, qui est encore la ville, pour gagner les bois, les forêts, les espaces lointains' (32). The facilitation of access to recreational spaces from homes and workplaces would enhance the daily lives of millions of suburban and urban residents.

One suggestion to help people move around cheaply and easily was the *métropolitain*, although Hénard had already pointed out its limitations. The *métro* was confined to Paris, and any extension beyond the city limits would open up more of the agglomeration to more residents, but was not politically popular at the time. Although many politicians were in favour of bridging the underground rail divide

33 When calculating the one-hour trajectory, Jaussely noted that, 'cette isochronie d'une heure comprend: le temps pour atteindre dans Paris le moyen de transport (dans le Paris d'aujourd'hui: gare, chemin de fer, tête de ligne de tramways; dans le Paris du début du XIXe siècle: point de départ des diligences), le temps employé pour parcourir la distance de ce point de départ au point d'arrivée dans l'agglomération de la banlieue, le reste de l'heure donnant la distance à parcourir pour se rendre de ce dernier point à son domicile' ('Chronique' 182).

between Paris and its suburbs, such a large endeavour would take time and much political capital. Between a major overhaul of the electrical capacity of the entire network and the successful passage of budget approvals, politicians from all sides would have to agree to any such proposal and where those extensions could occur. By 1920, the *métro* expansion discussion had reached an impasse. The city-suburb divide remained too strong for a quick resolution on this matter. The distinct physical and administrative separation between the city and suburbs also translated into little communication and no collaboration between the *métro* and regional rail companies. In reality, there were only two options for any possible suburban expansion: a comprehensive, costly extension at a time when the nation needed to rebuild after the First World War, or maintenance of the status quo with the possibility of a few minor extensions in the future. Political gridlock and lack of momentum ultimately dealt the former option a final blow and ended any hope of a large-scale expansion beyond the fortifications (Désabres 265). Even today, we see the result of this decision. The Régie autonome des transports parisiens (RATP) mainly serves the city and near suburbs, but no further. It would take the creation of a new entity, the Réseau express régional (RER) conceived in 1965, to step in and bridge this gap in access.

Paris still remained the hub for rail services, the principal means of long-distance travel in France, and with the exception of the Gare de l'Est, Jaussely kept Paris's existing train stations in their current locations. The map reader can easily find the red *gares* Saint-Lazare, du Nord, de Lyon, d'Austerlitz, and Montparnasse inside the yellow city. Commuters and voyagers already familiar with the city could access the same stations as before. These imposing structures anchored the 'voies de communication' that would feed development and expansion from central Paris into the suburban fabric. Bonnier emphasized their importance to Paris:

> Au cours du siècle dernier, les véritables entrées des villes ont été les gares, mais tout l'effort décoratif s'est porté à contre-sens, non pour séduire ceux qui arrivent, mais bien ceux qui partent. Ce n'est pas l'accueil de la cité, c'est la réclame dès la Compagnie. Si les façades des gares sont volontiers dites monumentales, leur envers et leurs accessoires sont piteux, pas étudiés et surtout malpropres. Les chemins de fer devant durer encore un certain

> temps qu'il est difficile d'évaluer exactement, il convient de nettoyer et d'embellir ces accès de la ville ainsi que ceux trop abandonnés qui correspondent aux grandes routes et que l'automobilisme ressuscite. (34)

Jaussely incorporated into his project an additional large rail station outside of the city, near Versailles, a prospect that Foncin also considered needed. His dotted red lines become solid black (maintained infrastructure per competition guidelines) as they join existing ones and branch out from the stations to the suburbs, provincial cities, and the rest of Europe. New red *gares de triage* in the suburbs, another enhancement, could help relieve some of the capital's rail congestion (Foncin 83). For practical, spatial, and quality reasons, Jaussely places these *gares* throughout the agglomeration and labels them according to their position (Gare de Triage du Nord, Gare de Triage de la Cie. de l'Est are two such examples). With even more oversight, the establishment of a rail hierarchy, dedicated tracks for long-haul passenger, freight, and suburban commuter trains could alleviate bottlenecks in and around the city. Some urbanists called for even greater change. A few Concours projects placed new rail stations where they could not effectively serve the public.[34] Jaussely was not against significant alteration when it could help the region. In one drastic change, he chose to move the Gare de l'Est from its home in the tenth *arrondissement* to a new location near La Chapelle (still within the city limits) next to a *gare des marchandises* near the La Villette compound. His other new warehouse stations throughout the region, such as one near Gennevilliers adjacent to the new port, followed the installation of said port, plus automobile and aviation factories (Lambert 105). Jaussely believed that supporting Paris's industrial base through improved infrastructure would help the region develop economically.

Both an extensive road network and industry would help grow the French economy. One of the most notable components is the large *port fluvial* to Paris's north-west, in present-day Villeneuve-la-Garenne.

34 Foncin critiqued a few projects in this regard. 'Crescabit nec mergabitur,' 'Sirius,' and 'Ex preterito futurum,' among others, suggested large-scale demolition of housing to create new stations. One entrant proposed a massive station for Palais-Royal, the Tuileries, or Les Halles. Foncin worried that such a project would crumble the cavernous 'sous-sol' eaten away by the métro, sewer, and other works (83).

This gigantic leaf-like addition next to the Seine would help vault Paris into competition with other major European port cities, such as Rotterdam, Antwerp, or Hamburg. Foncin devoted considerable discussion to the 'Urbs' port plan, and its accompanying canalization:

> Examinons par exemple son étude du port de Paris, si suggestives sur tant de points. Des travaux locaux destinées à faire de Paris un port régional sont d'abord envisagés (1). Dans le nombre, certains s'imposent comme le canal de dérivation de la Marne. Quelques-uns seraient utiles, comme le recoupage des méandres. Plusieurs sont moins indiqués comme le prolongement du canal de l'Ourcq dont le trafic est presque nul, la création d'un nouveau port du sud alors qu'on se demande si celui de Bonneuil, né de la guerre, pourra lui survivre. (81)[35]

Foncin wondered how much such an expensive, large-scale project would benefit the city and nation. With few natural resources to export, Paris was left with manufactured goods, which were better served by rail. Could or should Paris compete with already established ports elsewhere in Europe? In Foncin's mind, Paris should focus its resources on what it already had, which would improve the city.

One of the interesting aspects of Jaussely's map is his look into the future. Transportation improvements, by way of cars, trams, and trains, were important components to most plans, but Jaussely also tried to provide enhanced access to Paris for decades to come. Regularly scheduled air travel was not yet a reality for the majority of Parisians. The first commercial French airline, the Société générale des transports aériens (SGTA) was founded in 1919, not long after the publication of this plan and Le Bourget airport, to the north-east of

35 Foncin provided much detail in her decision: 'Beaucoup seraient difficilement exécutables dans une région déjà aussi peuplée que la banlieue nord. Par contre, l'aménagement du canal Saint-Martin n'est pas prévu. Cependant l'ensemble est presque acceptable. D'autres travaux, gigantesques ceux-ci ont pour but de créer à Paris un grand port national, international même. Il ne s'agit de rien moins que d'approfondir la Seine à 8 m. 50; d'établir d'immenses bassins à Achères et Gennevilliers; de relier par trois grands canaux Paris à Dieppe, à Angers (par Juvisy et Chartres) et Bordeaux (par Juvisy, Vendôme, le Poitou et les Charentes). Il est évident qu'ici l'auteur se laisse entrainer trop loin. Tous les ingénieurs, après une étude sérieuse du régime du fleuve, ont renoncé à approfondir la Seine à plus de 4 m 50' (81).

the city, started its business operations that same year. Jaussely's map recognizes the importance that this novel mode of transportation would have for Paris and includes a bright light-green *zone d'aviation commerciale région nord-est* to the north of a pink *zone industrielle*, curiously not at Le Bourget. Furthermore, Jaussely did not provide Paris with just one airport. He proposed a *zone d'aviation commerciale région ouest* near Versailles, adjacent to a *route nationale*, plus a *zone d'aviation commercial région sud* to the south of a large park, a *zone d'aviation commercial région sud-est* in the south-east corner of the map, a *zone d'aviation commercial région est* on the eastern edge of the metro area, and a *zone d'aviation Villacoublay* near Sceaux. Jaussely took the time to name each one on his document, even if only for their corresponding cardinal positions. In addition to their cartographic nomenclature, his non-natural colour choice helps them stand out on his map.

Through 'modern' additions such as aviation zones, Jaussely delighted his readers and allowed them to dream of the future. A city of this size and stature, Paris would surely need as many airports as it had train stations, if not more. Plus, the rise of military air forces during the First World War further demonstrated the importance of this new method of transportation. The well-known Michelin brothers, in addition to promoting tire use with the introduction and continued publication of the *Guide Michelin* each year, were early promoters of the French use of airplanes. In 1912, against a rising German threat to the east, they published a pamphlet entitled *Notre avenir est dans l'Air*, in which they called for France to acquire a minimum of 5,000 planes and pilots, plus the accompanying infrastructure (hangers, support staff, etc.) (Champeaux 61). If the French industry built planes to fight against Germany during the war, once the fighting stopped, they would build them for commercial purposes and Paris would be ready. The *zone aviation du sud* occupies a site near what is now Orly airport and has easy access via two *route nationales* as well.

Similar in colour to his airfields are two *autodromes*, car racing facilities. One of them, to the south of Versailles, also serves as a 'champ de manœuvre'; the second, named *autodrome* Spods-Divey, is to the north-east of Paris. These facilities complement the airfields in many ways. Not only do they suggest a growing comfort with new technologies among the bourgeoisie, but also changes in their use. Airplanes had other uses besides military and although automobiles changed how residents moved around the metropolitan area, in racing facilities, Jaussely provided an exhilarating recreational use for them as well.

Improved Hygiene

Roads helped people navigate the region. But the Commission made clear that entrants also had to help integrate Paris with the suburbs and Jaussely took pains to address the growing divide between the two. As new construction and residents enlarged Paris, the region's inevitable expansion had to be orderly. In his discussion of the entries to the Concours, Louis Bonnier pointed out a need to reduce the density of the city, a move that would make the urban fabric more manageable for residents and visitors. Plans were expected to 'crever les limites étroites qui font de Paris la plus encombrée des capitales, de renvoyer au plein air, à la nature, à la liberté, pour la majeure partie de son existence, le parisien anémié, gêné, étouffé, fébrile' (32). If planners created more space inside both the city and its inner-ring suburban communities, they could better balance population density and manage growth. The additional exposure to fresh air would improve the health and well-being of not only the working class, many of whom toiled in dirty, polluted sites, but the bourgeoisie in the city centre too. This philosophy mirrored that of Jules Siegfried and others at the Musée social.

Parisian by birth, Foncin, like many politicians, academics, and urbanists, disapproved of the capital's crowds and congestion.[36] For her, continued centralization on Paris at the expense of the rest of France represented the largest problem. Not only did too many people overrun the region's infrastructure, but trying to accommodate so many residents into a small area was not healthy, it would eventually inhibit economic growth and even become 'une cause de mort' for France (80). To reduce overcrowding, the Commission hoped to receive proposals that reserved open spaces, addressed construction quality, distribution of drinking water, sewers, removal of refuse,

36 Albert Sarraut, Minister of the Interior, lamented the fact that 'chaque usine qui sort de la terre appelle des centaines, sinon des milliers, d'ouvriers' (*Compte rendu* 2). See Chapter 5.

and land clean up ('Le Programme et le résultat du concours' 91). This wish list, albeit lengthy, established the direction which modern cities must take. A dirty, crowded, congested agglomeration would eventually collapse under a growing cascade of problems. Therefore, to ensure Paris's future, substantial improvements were needed to reduce crowding.

The idea of cleaning up Paris was not a new concept. Haussmann had spent time and money transforming the city's central medieval core into a showcase of nineteenth-century modernity. Beyond the new boulevards, he cleared out the slums and built parks and government buildings. These new bourgeois sections of the city were a success. However, even though his successor Alphand had spent time and energy in the outer *arrondissements*, neighbourhoods there remained unhealthy and in need of work. The Commission also felt that most suburbs too lacked proper attention. Albert Sarraut compared the development in some of the *cités suburbaines* to 'des cellules cancéreuses' (*Compte rendu* 1). Housing availability and quality beyond the wall had become such a problem that the Commission members believed it would require a large-scale solution, furthering the need for a regional plan. Technology had altered how people experienced the city and planners needed to adapt. For the area to become less dense, leaders needed to include more openness in the design. Improved access to green spaces and fresh air was of course a major point of the Musée social's political and social agenda. As planners looked to organize the city, these ideas became a precursor to zoning. In his 'Chronique de l'Urbanisme,' Jaussely brought up the need for a specialization in land use, which would create a more harmonious and organized urban fabric: 'La création de parcs, de jardins publics, celle de quartiers spéciaux de circulation ou d'affaires, d'administration, de jeux et de sports, l'établissement d'habitations à bon marché, etc., etc., jusqu'à présent entravés par des difficultés d'expropriation seraient facilités, rendus possible' (194). Most would agree that additional parks alone would change the regional dynamic and open up the city more, but that segregating industrial activity away on the periphery, away from homes and parks, could provide benefits as well.

Multiple Parisian neighbourhoods, especially in the eastern half, hosted both industry and slum housing. Although Haussmann cleared many of both from central Paris during the Second Empire, areas of north-eastern Paris, especially near La Villette, and south-eastern Paris, near the border of Ivry, contained many factories, many of which polluted water, land, and air. On the edge of the city's nineteenth *arrondissement*, one such example, the city had installed a *gazomètre* (a natural gas storage and processing facility) between the *chemin de fer de la ceinture* and the fortifications. This industrial installation, near the abattoirs and the Canal Saint-Denis, but also close to the Église Saint-Denis and undoubtedly to homes, would not have been a pleasant neighbour.

To the south of Bercy in the twelfth *arrondissement*, a *manufacture de tabacs* existed near a cemetery and the *gare aux marchandises* serving the Gare de Lyon. This industrial quadrant on the city's south-eastern edge contained both the Église Notre-Dame-de-la-Nativité (near Rue de la Nativité and Rue Liberté) and an *école des filles*, which suggested that Parisians lived nearby. Schoolgirls and churchgoers are not a combination one associates with warehouses and cigarette manufacturers. In the city's south-western fifteenth *arrondissement*, Paris had an *usine à gaz* on the Rue de la Croix-Nivert, across the street from a Protestant church and its accompanying *école protestante* and near the Place du Commerce. Louis Bonnier recognized how ill-fated combinations such as these, present throughout many neighbourhoods in the region, were bad for residents. He called for 'la répartition équilibrée des quartiers industriels, commerciaux, résidentiels' in order to create 'l'esthétique totale' of the Paris region (31). The various salmon-coloured *zones industrielles* marked on Jaussely's map further represent reformers' influence in French urban planning. This colour choice reminds one of grimy air. One can see how unhealthy they are. By placing these zones, and their residual pollution, away from residential neighbourhoods in the city of Paris, and the suburbs, Jaussely would improve the quality of life for the people who lived there. One of the largest of these zones, located to the north-east of La Villette, suggests that Jaussely agreed with Bonnier and would keep this sector of the agglomeration as an economic engine, even if it was a dirty one. He even labelled a sector in the *zone industrielle* 'zone des industries insalubres,' which further affirms their unclean nature. To the city's south, past Ivry, a second large industrial zone can be found, and a third to the north-east of La Défense. Jaussely placed these dirty, polluting complexes far from the embellished capital and its residents, but near to transportation so that workers could access them. New roads traverse and border the industrial parks and would facilitate commerce and the

French economy. The slight repositioning of Les Halles to La Villette, with its easily visible dark-red *marché aux bestiaux* and abattoirs to the north-east in a larger cartographic and physical space, further confirms their importance along with Paris's ability to feed itself, and maintains that corner of the city as a continued industrial site.

Thanks to his work with the Musée social, Jaussely hoped the government would work to improve the health and hygiene of the region's residents. Better access to healthcare would further demonstrate that the republic continued to address the needs of its citizens. Jaussely included many new hospitals throughout the region and painted each 'Hôpital général' with a dark-green square and a red cross. One near the Grand Parc du Sud, one along the Route Promenade Paris-Front past the Grand Parc de l'Est, on west of Boulogne, and one between the *zone industrielle* adjacent to La Défense and the *port fluvial*. These medical centres suggest where Jaussely imagined many new residents could move around. Perhaps most interesting, Jaussely suggested the construction of nine new hospitals in the ring of parks that would succeed the fortifications. Although the Musée social had recommended no building in that zone (for health reasons), Jaussely chose to go against that dictate.

Not only did he take care of the living residents in Greater Paris, Jaussely made sure to take care of the dead as well, as he included various necropolises, giant cemeteries outside the Paris city limits. Under Haussmann, officials had worried about the overcrowding of Paris's 13 cemeteries inside the wall. In the Third Republic, the Seine prefecture took steps to increase cemetery capacity in the suburbs and opened or expanded cemeteries in Saint-Ouen (1872), Ivry (1874), and Clichy (1883). In 1884, government officials chose the village of Pantin, just beyond La Villette, to house a new large cemetery. Its location, between two large roads just 5 kilometres from central Paris, meant that people would be able to access it easily from the city. This decision to place a cemetery within a commune highlighted the larger debate of eminent domain and how much say local leaders had in what was built within their boundaries.[37] The Pantin municipal council fought the dictate, citing 'les conséquences et le dommage' which the cemetery's construction would cause the village. Council members turned to the

37 The 1906 loi Bauquier gave towns and villages more rights in regard to what could be built or placed in their communities.

media, elected deputies, senators, and Paris city councillors to fight the Seine prefecture's move. In the end, however, they could not stop the process and the cemetery was built (Fourcaut et al. 105). Having set a precedent in Pantin, the Seine *préfet* next built cemeteries in Bagneux in 1886 and in Thiais in 1920, after the publication of Jaussely's plan. Jaussely saw this phenomenon as a way to both provide more building space in the city and compartmentalize land use. His dark blue necropolises, including Pantin's, are spread throughout the agglomeration. This addition to Parisian urban space highlights the region's ability to take care of the deceased in an orderly manner.

Clean water rivalled clean air when it came to providing a healthy living environment for the region's citizens. Throughout the second half of the nineteenth century, engineers under Haussmann worked to provide Paris with an up-to-date sewer system. Unfortunately, the refuse water collected under the streets emptied out into the Seine via three large collection points. In order to begin treating the growing amount of wastewater produced by the city, the government passed the loi du 4 avril 1889, which focused on the purification of the Seine. The construction of the Achères water treatment plant, one result of this law, provided fresh water and showcased French engineering prowess. Processing pools could be found where the Seine and the Oise rivers come together, in 1896 the government eventually expanded the project and built locations near Triel and Méry-sur-Oise. To enhance their value to residents, Jaussely afforded considerable cartographic space to these treatment plants on the banks of the Seine. He wished to enhance the city's ability to treat its wastewater and proposed several light-lavender 'Épuration biologique des eaux usées,' primarily near the *zones industrielles*. This bright colour differentiated them from the 'dirty' factories and affirmed that Paris could improve land and water. There is a large one near Poissy and another north of the large salmon-pink *zone industrielle* north-east of the city. Also, next those two zones, Jaussely included a bright-lavender 'Incinération OM' (Incinération des ordures ménagères), which would incinerate and eliminate the region's trash. Two other 'Incinération(s) OM' stand alone – one near Versailles and one next to the *zone industrielle* by La Défense. With land use regulations, new cemeteries away from the crowded city centre, and sewage treatment plants, Jaussely's plan provided Paris with healthier air, water, and land.

The Urban Aesthetic: More Green

As officials improved the overall environment of the Paris region, Commission members also hoped to improve the area's aesthetic. Much time had gone into planning how to improve hygiene. There is a significant amount of green on Jaussely's map, which conjures up images of leafy parks. His twentieth-century readers would expect to see parks, more so than sewage treatment plants or even *aérogares*. The creation of more parks would improve residents' health but also change how the Paris region looked and felt. Jaussely's map is full of green and the proliferation of this colour further suggests the Musée social's influence. Perhaps of interest to most urbanists are the nine parks in the fortification ring proposed by Hénard and used in the Musée social's 1908 municipal election campaign. Beyond the hospitals mentioned earlier, Jaussely planned other building there. Instead of Hénard's Parc de la Villette, as noted, Jaussely suggested moving Haussmann's crowded Les Halles Centrales from the centre to this corner of the city. In the suburbs beyond the fortifications, however, Jaussely proposed significant access to parks and forests for residents in all quadrants. To the south of the city, Jaussely suggested six parks on the left bank, none of which were actually built. One of the largest, the Grand Parc du Sud, fits snuggly between two large roads heading towards Orléans. Roughly the same size as the Bois de Vincennes to the east, this large southern park could have provided significant open space to the Kremlin-Bicêtre area. To the east, Jaussely incorporated a large green band, which would be named the Grand Parc de l'Est. This greenway departs from the fortifications, near the Porte de Lilas, goes round the suburb of Montreuil, down towards the Bois de Vincennes. To the north-east, adjacent to a large *zone industrielle*, another large green space, the Parc de Mont-Mesly, would help cleanse the air polluted by the nearby factories.

Various *cités-jardins* could be found throughout suburban Paris. This new idyllic settlement type: suburban communities with individual homes and gardens represented an idealistic solution to urbanists around the world. The loi du 18 juillet 1915 officially created the Office départemental d'habitations à bon marché du département de la Seine, which coordinated these suburban developments. That legislation called for the Seine departmental HBM office to work with its Parisian counterpart and others in the region to intervene on behalf of the working poor and build affordable housing. Consolidated, then renamed in 1918 to Cités-jardins du Grand Paris, the agency was directed by Henri Sellier, the socialist mayor of Suresnes, for more than 20 years. The goal of the bureau was to create residential suburban communities and link them with Paris via rail or tramway lines. Potential *cités-jardins*, for example, would complement the existing villages and communities along the various valleys and depressions mentioned by Foncin. At the time of the 1919 loi Cornudet, various individuals proposed the creation of *cités-jardins* throughout Paris's suburban ring. Engineer Raoul Dautry and architect-urbanist Jean-Michel Auburtin planned an important *cité-jardin* to Paris's north, near La Courneuve, Le Bourget, and Dugney. However, the project eventually fell through due to lack of transportation to that site and so the Dautry/Auburtin project receives no cartographic space on Jaussely's 1919 plan. Paul de Rutté, Joseph Bassompierre-Seburin, and Paul Sirvin, all associated with the architect Maurice Payret-Dortail, who worked for the city of Paris and the Office Public d'HBM, proposed the creation of a large *cité-jardin du Grand Paris* to be linked with the city upon the electrification of the Luxembourg–Limours rail line in Châteney near Sceaux, though that too is not included on this map. But many other *cités-jardins* are, and they can be found in every corner of the metropolitan area. Their light pale-yellow colour, combined with their large number, almost normalizes them. It suggests that Jaussely believed that they would be the norm throughout the suburbs. He carpeted the area with solid residential construction surrounded by lawns, gardens, and relatively adjacent large parks and woods. One particular *cité-jardin* was also to be part of a principal *campus universitaire* (Sellier 53). Jaussely's map included that *cité-universitaire* not far from the Grand Parc du Sud, traversed by a wide avenue which would become the A86.

The idea of moving university housing from central Paris into the suburbs arose with some frequency in early twentieth-century French urban debates. Not only did this discussion centre on providing high-quality affordable housing for French students, it would also help develop the suburbs and help expand educational opportunities there. Foncin noted that most plan contributors felt that there should be some sort of transfer not only of student residents, but also of portions of the centre of learning from the ever-crowded Latin Quarter to the region near Sceaux, a site chosen due to its relative lack of development compared to other suburban communities and room for growth

(Lambert 104). The construction of a new 'cité-jardin intellectuel' would allow the government to cultivate laboratories and classrooms alongside healthy students, trees, and flowers. Foncin referenced Cambridge and Oxford in England and Harvard in the United States as possible educational models for the French system (87–88). However, she actually came out against the move and identified problems that would arise from a decision of this magnitude. The physical separation from the city would require enhanced transportation during and after the transition. The enormous cost of construction and the potential disruption to the neighbourhood surrounding the present-day Sorbonne campus also weighed heavily against such a move in her mind. In addition, any campus transfer would isolate professors and students from other venerable institutions and the city's intellectual heart, such as the Collège de France, museums, archives, and theatres (88–89). Although others, such as Foncin, argued against such a move, Jaussely included a *université* adjacent to his *cité universitaire* site. They had little to fear, however. As demonstrated by other improvements, change happened slowly. On 1 January 1971, almost 50 years after Jaussely published his plan, the government opened the Paris Sud XI's Faculté Jean Monnet campus nearby, as part of the post-1968 higher education reforms.

Housing was a prime concern, but officials also hoped to see how planners would improve the quality of life of the region's residents in other ways. In terms of education, a cornerstone of Third Republic values, Commission members wanted to see the construction of new schools and the replacement of old, outdated ones. New hospitals had to be built in the suburbs. Les Halles Centrales, the belly of Paris, needed to be enlarged – and perhaps moved – and similar establishments placed throughout the suburban ring. Trash collection and waste management, a concern in the large urban area, needed to be addressed. To support that growth, new electric and gas power stations would help create a reliable supply of power for the city's industry.

No Plan for Paris

On 13 May 1924 Parisians awoke to the front page of *L'Humanité* declaring that, 'Les ouvriers de la région parisienne ont voté pour la Révolution [...] en banlieue nous battons tous les blocs bourgeois.' This triumphant headline announced the Cartel des gauches's large-scale first-round electoral victory in suburban Paris. *L'Humanité*' then proudly declared the precarious state of the bourgeois city: 'Paris, capital du capitalisme, est encerclé par un prolétariat qui prend conscience de sa force.' Although the Commune-esque physical combat witnessed suggested by the vocabulary was far from likely at this point, victory here demonstrates the suburbs' desire to assert control over their own affairs, even at the expense of government wishes. The political victory that created a *ceinture rouge* outside of Paris brought the working class into the national political discussion for the first time. Even if some Parisians awoke on that May morning surprised at the size of the victory just outside the city limits, Jaussely's *plan d'extension* presented the importance of the Paris suburbs.

The 1919 loi Cornudet proudly introduced the idea of French urbanism to the nation. In reality, however, Jaussely's *plan d'extension* represents more of an ideal and a true blueprint of what a modern French city should look like in the future, than an actual usable plan for urban reform. Little change actually happened. As late as 1924, five years after Jaussely presented his map, François Latour, Rapporteur général du budget de la Ville de Paris et du budget du Département du conseil général de la Seine, worried about controlling the chaotic suburban construction outside the city. Latour saw little to no improvement from the 1919 legislation and felt that there was too much resistance to politicians' ideas in the suburbs. The situation needed to change. He doubted that 'la capitale l'élégance et du goût puisse impunément rester enfouie au milieu des lotissements sordides et de la zone pestilentielle' (*Banlieue* 172). Paris seemed trapped under the spectre of Haussmann's top-down shadow, even 50 years later, which continuously slowed down any improvement to the city.

Beyond the stark Paris-suburb divide, the difficulty of his proposals, and antipathy towards any meaningful urban merger, other problems hindered the implementation of Léon Jaussely's 1919 *plan d'extension*. Eugène Hénard, Jaussely's mentor and an influential and strong personality, was ill and in retirement by the time the loi Cornudet passed. Although not in the competition himself, Hénard's voice was especially apparent in Jaussely's plan. But Jaussely lacked Hénard's name recognition and influence and the cost of post-war rebuilding compounded the government's lack of political will to continue. Add to these limitations the push back from the *banlieues* and the chances of Paris constructing the infrastructure found in the *plan d'extension* seemed remote.

Some faulted Jaussely's plan itself. Although some critics praised his knowledge of the terrain, others said his project proved to be so hastily designed and so insufficiently documented that it was totally unusable (Wolf 90). More practically, political shifts changed how departments and cities applied Cornudet's law. Intercommunal cooperation, one cornerstone of the legislation, fell sharply throughout the 1920s as locally elected politicians pushed community-centred improvements at the expense of region-wide ones (Gaudin 153). Furthermore, in Paris as in other large agglomerations, communities terminated agreements creating region-wide cooperation as envisioned under the loi Cornudet. Even the creation of a Commission supérieure d'aménagement in Paris, the one regional organization that was to plan for the department, had a troubled start. Its two principal activities, a preliminary council on urban affairs and a technical advisor for new projects in the region, proved problematic. Although the two roles complemented each other, their announcement in the 5 March 1920 Cornudet circular came out in vague and hypocritical terms (Gaudin 164). These problems thwarted the loi Cornudet and lessened the chances that Paris would fully implement Jaussely's *plan d'extension*. In 1927, Henri Chardon, Président de Section au Conseil d'Etat, expressed his frustration with the end result: 'En 1919 (dès la cessation de la guerre), le plan [du loi Cornuet] fut mis au concours. Une abondante moisson d'idées en résulta. Notre confère Jaussely obtint le premier prix. [...] Puis [...] ce fut le silence absolu!' (2). Ten years later, politicians remained perplexed by the failure to reform the suburbs.

Regardless, although never implemented as designed, Jaussely's cartography illustrates the changing notion of cities in France at the beginning of the twentieth century. Furthermore, Jaussely's ideas demonstrate the influence of urbanism, a new study of cities, on the Paris region and his desire to plan region-wide improvements. The growth of parks would now complement road infrastructure. The prominence of the *banlieue* on the *plan d'extension* demonstrated the need to incorporate new development beyond the wall into the greater urban conversation. Jaussely's map also reveals how the idea of Paris, no longer simply the city itself, has changed. When the French thought of the capital, it was much more than the land inside the wall. Automobilists and rail passengers alike traversed built-up suburbs as they entered Paris. Residents who were pushed out by Haussmann's transformations and new immigrants from the rest of France populated the large ring. The new science of urbanism influenced Jaussely and other French urban planners. This science encouraged city leaders to plan for the future and improve the lives of many, based in science. Access to fresh air and clean water can make happier, healthier residents. As the automobile took hold, Jaussely continued to improve the city's infrastructure.

Jaussely employed the army's 1908 *état-major* map as a base upon which he painted the improvements and changes he wished to make to the Paris region. Jaussely affirmed his position on Thiers's fortifications, subject to intense debate. He 'removed' them from the city, and its future, in one easy step by simply covering them with a coat of light-green paint. Nine green parks proposed by Eugène Hénard are absent and the only buildings the reader can see are a few construction projects, such as a new Les Halles near La Villette. Considering that the fortifications question was so central to the Musée social's mission, Jaussely confirmed their plan. By concealing them from the readers, he has in effect dismantled them for the city. Paris, under his *plan d'extension*, would finally get a ring of open space and fresh air. By the mid- to late 1920s, however, change was still needed.

Chapter 6

The Rise of Suburban Paris

Henri Prost's *Carte générale*

> Cette œuvre est vôtre. C'est vous qui l'avez imaginée, c'est vous qui l'avez créée, c'est vous qui l'avez mise en train, en instituant notre comité, il y a six ans, le 28 mars 1928. Par une sorte de justice distributive, bien rare, en politique, c'est à vous que revient l'honneur mais aussi la lourde tâche de la poursuivre. Si, grâce à votre haute autorité, les Chambres réalisent toute la portée et toutes les conséquences de la loi qu'elles ont votée en 1932, elles sauront la compléter et en assurer jusqu'au bout l'exécution. Le comité supérieur, fidèle à son mandat, leur en apporte par votre intermédiaire, dans les délais légaux, les moyens techniques et juridiques. Au Parlement de faire le reste. Vous aurez ainsi, monsieur le président, attaché à jamais votre nom à la plus grande entreprise d'urbanisme national.
>
> —Louis Dausset président du comité supérieur de l'aménagement et de l'organisation général de la région parisienne à M. Albert Sarraut, Président du Conseil des ministre français, Ministre de l'Intérieur, le 15 mai 1934[1]

One Sunday afternoon in 1928, Raymond Poincaré, the Président du Conseil des Ministres (Prime Minister), returned from a weekend at his property at Sampigny in the Lorraine region. The weather was beautiful and the trip was going smoothly until he arrived at

1 Dausset's discourse, accompanied by Henri Prost as he presented his plan to Albert Sarraut, can be found on the front page of *Le Petit Parisien* from 15 May 1934 with a photograph of the event.

Chelles, on the eastern outskirts of Paris. There, he found himself behind a long line of cars. An automobile accident ahead had closed a portion of the highway and Poincaré, unable to move, encountered first-hand the realities of unregulated Parisian sprawl. From his car window, he witnessed, 'une banlieue interminable, sans ossature, où le hasard seul semblait offrir un passage, [avec des] habitations noyées dans un indescriptible chaos' (Remaury 146). The narrow streets and substandard construction, reminiscent of pre-Haussmann Paris, seemed to startle him. Once back in the city, moved by what he saw, Poincaré suggested a formal region-wide planning authority to normalize not only suburban housing construction, but also the increasingly over-burdened road network outside Paris's city limits.

Six years later, on 14 May 1934, French urbanist Henri Prost presented Albert Sarraut, Minister of the Interior, with his 'grand album contenant le projet d'aménagement de la région parisienne (Plan Prost),' a dossier with his detailed plan for Paris's future (Remaury 152). A press photo of the event shows Prost, Dausset and Sarraut, along with prefects of the Seine, Seine-et-Oise, Seine-et-Marne, and Oise departments, standing over and discussing the newly created 'plan de réaménagement de la région parisienne.' Prost created his plans to improve access to the city, decongest the suburban ring, and assure the conservation of open space and woods in the region. Beyond a written proposal, his project also included multiple illustrations of various *sorties*, exit points to the provinces, and a *Carte générale de l'aménagement de la Région parisienne*, a brightly coloured, multipage representation of Paris and its suburbs. Prost's map, the last officially approved cartographic proposal for the capital published under the Third Republic, recognized the changing nature of early twentieth-century cities, where the automobile enhanced personal movement but had a tendency to overwhelm nineteenth-century infrastructure.[2] As part of the urban planning process, the Comité supérieur de l'aménagement de la région parisienne, Paris's regional planning authority, had asked potential planning candidates, including Prost, to 'concrétiser tout d'abord les limites de la région parisienne, en fonction des grands courants généraux de circulation' (Remaury 152). Prost's *Carte générale*

2 Examples of such plans were the Commission d'extension de Paris's 1913 *Avant-projet d'un plan d'extension de Paris* and Léon Jaussely's 1919 *Projet lauréat de la section générale du concours du plan d'extension de Paris*.

reinforces the desire to both know and control the growing region and addresses current transportation infrastructural inadequacies. The map also prominently highlights new autoroutes and divided highways, clearly delineates – geographically – where the region ends officially, and authenticates Paris's newly enlarged shape. Prost acknowledges the city's changing suburban identity and recognizes the growing presence of the *banlieue*. Perhaps most important, the Plan Prost attempted to remove the suburban unknown. His maps document not only future development and existing green space, but also suggest how to control urban growth. This chapter will investigate how Henri Prost's *Carte générale* demonstrates the French government's desire to move beyond the ideals of urbanism, as found in Jaussely's 1919 *Plan d'extension*, and bring them to fruition. Although Jaussely had proposed new and exciting ideas for Greater Paris, Prost provides a much more realistic plan to address the growing region's needs. In addition, his presentation of the suburban belt, at times both feared and hated, helped introduce it to many Parisian residents. Most importantly, Prost's *Carte générale* demonstrates what is possible when large-scale infrastructural improvements are able to traverse the whole region.

Prost's *Carte générale*

Hoping to end unregulated Parisian sprawl, the government looked for a plan to structure and to control the Paris region. To do so, urban planners and cartographers needed to acknowledge the difference between 'la vue perspective,' which looks back at what a city has accomplished and 'la vue prospective,' which may involve 'un futur incertain' (De Certeau 142–43). The Musée social movement and Jaussely helped France turn away from the perspective viewpoint, where cartographers documented what existed in a city at a certain point in time, which had been an unofficial norm in official French cartography. Thanks to these new social and technological influences, planners such as Prost could focus not on the past, but on solving problems. As he presents the well-expanded metropolitan area in significant detail, Prost affirms that the suburbs should no longer remain 'unknown' to city residents and officials, and invites towns and villages outside Paris into the greater urban conversation. Through colour, he assigns value to regulated housing, industrial infrastructure,

Figure 7. Prost, Henri, 'Carte générale de l'aménagement de la Région parisienne, région est et ouest,' 1934.

green spaces outside Paris's city limits as well as the comprehensive road network. Furthermore, he foretells the future of the region as the Comité supérieur worked to merge the urban centre and multiple suburban *communes* into one political entity. To emphasize his intentions for the suburbs, their colour presentation monopolizes the map and provides the *extra-muros* territory with never-before-seen cartographic prominence and power.

Taking a look at Prost's expansive *Carte générale*, the 'Plan du projet directeur approuvé' comes in three over-sized sections: a 'région nord,' a 'région est et ouest,' and a 'région sud.' Due to the sheer scale of the Paris Basin, Prost did not attempt to fit his project on one single sheet, a decision that also lent credibility to the extent and complexity of his plan and the amount of work the region needed. Reminiscent of de

Certeau's view above New York, Prost's three maps provide a global survey of the land around Paris, the scale of which is matched by their significance for the nation. This is the first map produced for Paris with the explicit goal of planning and ordering the chaotic region, not one that provided just a wish list, like the *Plan d'extension* published by Jaussely.

Cities or metropolitan areas are comprised of a series of competing relationships around urban nodes of production or transportation with the strongest areas, for example city centres, becoming the most developed and receiving the most resources. Geographically, 'Paris' had grown in the 15 years since Jaussely's 1919 *Plan d'extension* and Prost's *Carte générale* reflects this fact. It represents the first detailed investigation on a regional level. A technical product of Prost's urbanist background, the overall plan is an example of 1930s French urbanism and foretells what other urban plans should look like in the future (Laisney 120). Through his map, readers can better visualize Greater

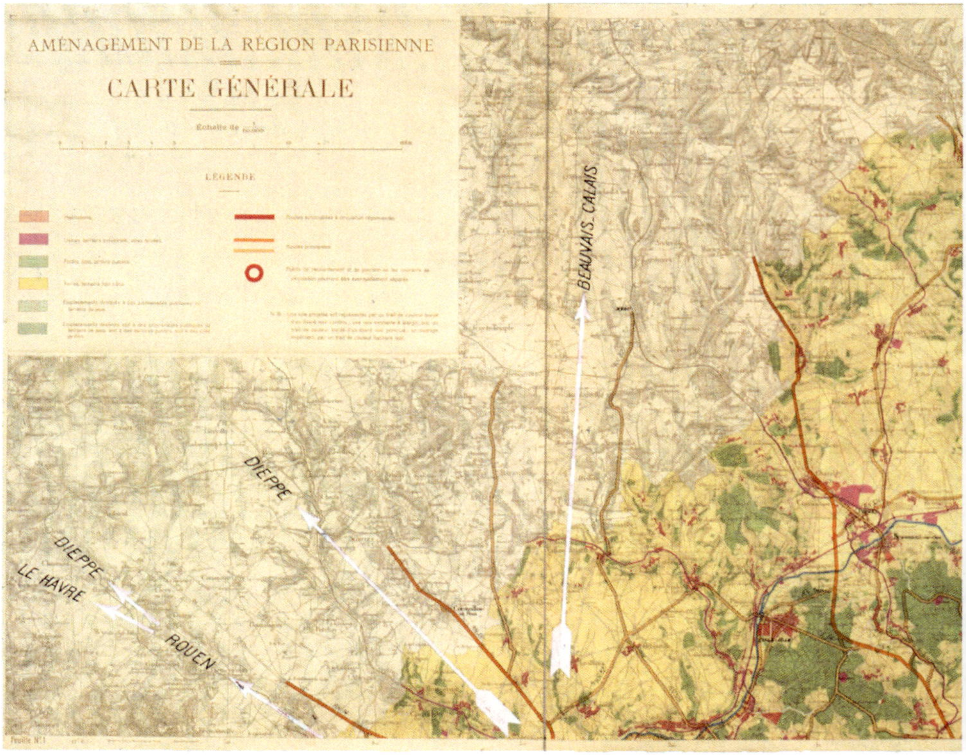

Figure 8. Prost, Henri, 'Carte générale de l'aménagement de la Région parisienne, région nord,' 1934.

Paris's continually changing hierarchy and complexity. Furthermore, the *Carte Générale* illustrates where the region must react to unregulated suburban development, population growth, and automobile use in the first half of the twentieth century. When it came time to create the cartographic documents required by the 1932 legislation, Prost turned to the personnel at the Chemins de fer de l'état to reference their existing cartographic surveys. Similar to Jaussely, the rail company employed the army's *Carte d'état-major* and, with the Service géographique's document as a base, Prost had in his possession the preliminary tools to publish his maps in time for the two-year deadline.

Presenting the Suburbs

Starting in the late eighteenth and continuing into the nineteenth and twentieth century, relations between European cities, including Paris, and their suburbs had transitioned from 'a hierarchical to a contrapositional structure. Rather than the external communities being merely subordinate to cities, they came to be seen as oppositional' (Dennis 179). By the 1920s, Paris's sizeable industrial belt – with its factories, transportation infrastructure, shanty towns, and numerous former rural villages swallowed up by unchecked growth – absorbed thousands of new residents each year. Home to much of the region's working-class population, this unregulated zone marked a sharp divide from Haussmann's embellished city centre. Cheap land and few regulations fed the physical and mental division between the two opposing sides.

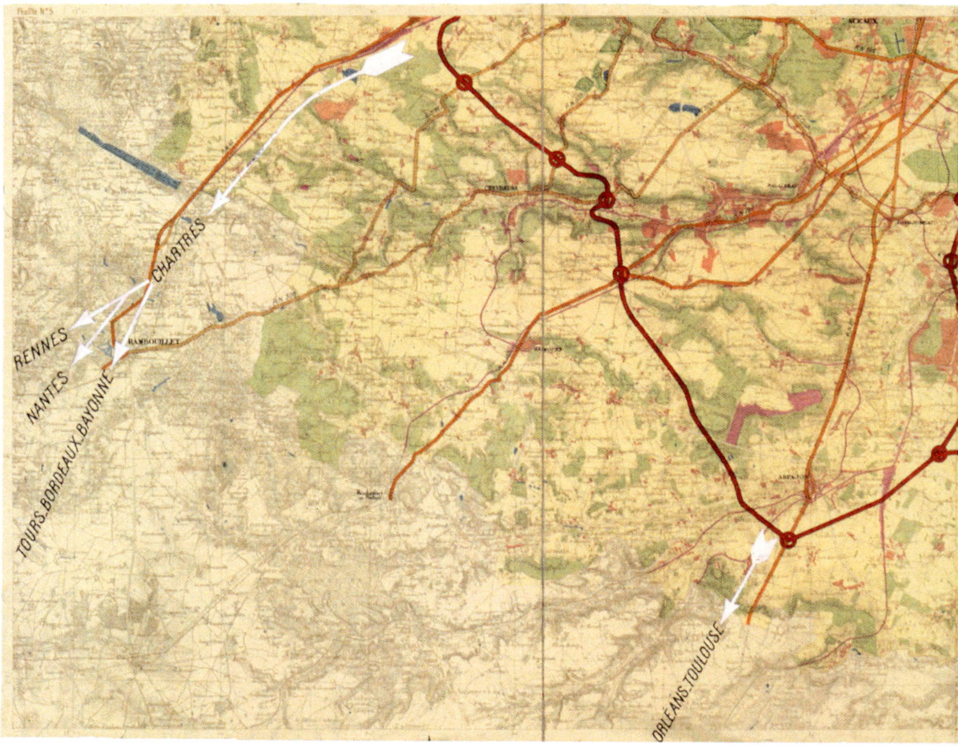

Figure 9. Prost, Henri, 'Carte générale de l'aménagement de la Région parisienne, région sud,' 1934.

Even after the passage of 1919 loi Cornudet and Jaussely's subsequent *Plan d'extension*, Prime Minister Poincaré's 1928 experience demonstrated how leaders continued to downplay the importance of, or outright ignore, the suburbs. With little oversight, unregulated blight affected not only the ability to enter and exit the city easily, but also created a sizeable working-class political enclave in the industrial suburbs just outside Paris. Residents there delivered significant electoral victories in 1924 to the 'cartel des gauches,' a mixture of leftist parties including the Radical-Socialists and the SFIO (Section française de l'internationale ouvrière), and gave the territory the *ceinture rouge* label. The success of the elections on 11 May that year came about in part due to the power of the revolutionary proletariat's control over the reactionary suburban neighbourhoods (Marchand, *Les Ennemis* 114).

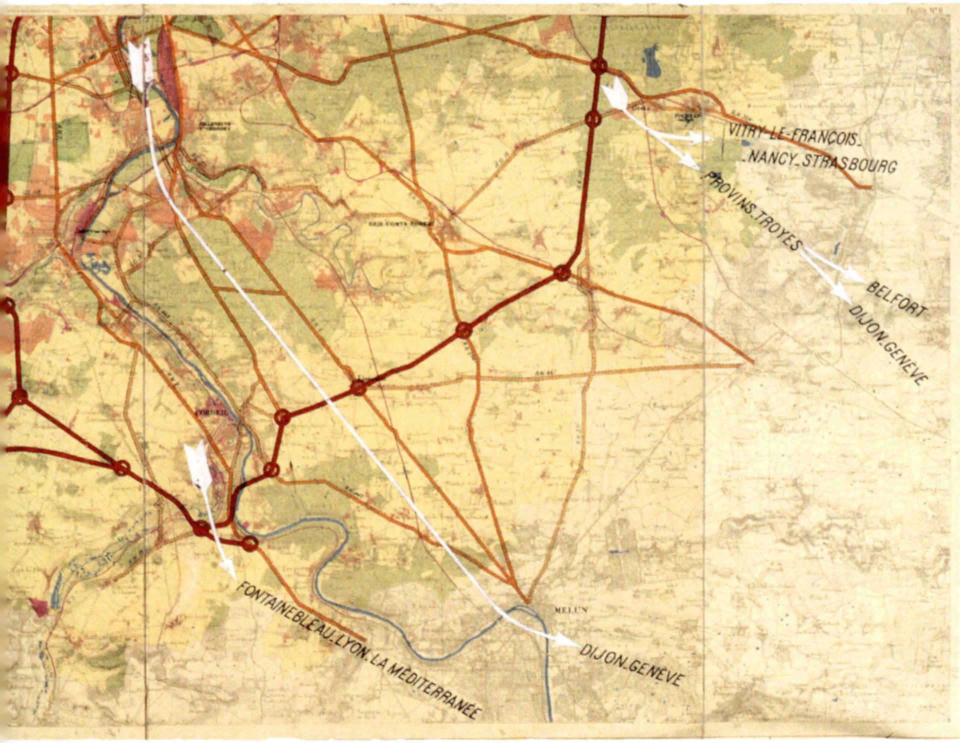

Beyond the suburbanites themselves, bourgeois leaders also wished to understand and control the building and growth in Paris's first ring of suburban villages and towns, since the Seine department administered both the city and surrounding *communes*.

Each *commune* was in charge of its own planning, housing, and construction; a situation that exaggerated the disparate nature of the region's growth. Some municipalities, especially those adjacent to the capital, often followed Paris's regulations (Laisney 119). Others, however, had neither the means nor the desire to duplicate the capital's decisions. The lack of continuity in such a densely populated area, combined with no practical ability to enforce regional directives, often meant that two adjacent communities could have misaligned regulations on building or road construction and little desire or incentive to cooperate, even for the greater good of their residents.

The new science of urbanism, formed in the early twentieth century, influenced heavily by the Musée social, and discussed in the preceding chapter, meant to encourage greater regional collaboration. Although it initially concentrated on the improvement of hygiene, the movement also studied how cities grow and what path they should take in order to enhance the lives of all residents:

> Dans tous les pays où l'on a fait de l'urbanisme, où l'on a fait une véritable extension des villes, on a voulu décongestionner les grandes agglomérations urbaines pour leur donner de la lumière, dans tous ces pays on a bien été obligé d'établir des servitudes et de créer des taxes sur les plus-values, afin que chaque bénéficiaire de travaux publics puisse contribuer à ces travaux dans une certain mesure. (Dausset xix)

The rise of French urbanists, such as Hénard, Prost, Jaussely, Bonnier, and Benoît-Lévy brought forth the idea that larger comprehensive plans would enhance the movement of people and goods, raise land values, and harmonize political will across economic and political factions in large metropolitan areas. These projects could also improve the lives of the millions who lived in them. But could such a plan work for Paris? Albert Sarraut, Ministre de l'Intérieur, spoke of French limitations:

> Notre législation, notre législation communale surtout, est trop souvent conçue d'un point de vue statique, pas assez dynamique. La croissance des villes y apparaît comme une sorte d'anomalie. Il a fallu attendre l'année 1919 pour avoir en France une loi d'urbanisme; encore cette loi n'est-elle pas toujours exactement appliquée. (*Compte rendu* 2)

The unruly nature of suburban Paris made any attempt to tame it difficult. Following the First World War, the government had tried, and failed, to implement Jaussely's 1919 *Plan d'extension* for the region. Any new effort now would require support from Paris and would need to focus on the array of individual communities throughout the vast suburban belt.

Even if politicians wanted to continue to prioritize Parisians over suburban residents, as they had throughout the first two decades of the twentieth century, the region's demographics suggested that such a move was unpractical. In his *Avenir de Paris* (1925), urban critic Albert Guérard described the growth of the suburban departments around the nation's capital. Paris's 2.9 million residents, of the 4.4 million of the Seine department, affirmed the city's demographic advantage in the region, but in the first half of the 1920s, the capital actually lost people – around 35,000 residents – while the rest of the department gained 217,000:

> L'accroissement de l'agglomération, qui n'a rien d'anormal, s'est porté tout entière sur la banlieue; il serait plus frappant encore si l'on faisait entrer en ligne de compte certains districts de Seine-et-Oise. Si ce double mouvement continue, dans quelques années, Paris ne sera guère que la moitié de Paris. Les fortifications de M. Thiers et l'annexion de 1860 ne correspondent plus le moins du monde à la réalité. (366)

This population increase resulted primarily from the steady migration from the provinces for jobs in the nation's growing industrial sector and many of these new residents settled in the suburbs, not in the city.

In 1928, Albert Sarraut too spoke of French internal resettlement:

> Les causes de cette migration, nous les connaissons. Elles résident, en premier lieu, dans les progrès de l'industrie. Chaque usine qui sort de terre appelle des centaines, sinon des milliers d'ouvriers et je ne pense pas que les industriels se préoccupent toujours de loger dans de bonnes conditions, ni même de loger, absolument parlant, les familles de ces travailleurs. (*Compte rendu* 2)

Henri Chardon, Président de Section au Conseil d'Etat, echoed the Interior Minister's sentiments. The draw of Paris's industrial concentration, which started in the nineteenth century, continued unabated into the 1920s and 1930s. In his 1927 address to the Académie des Sciences Morales et Politiques, Chardon mentioned a conversation he had had with an automobile manufacturer. 'Pourquoi ne cherchez-vous pas à installer vos ateliers près d'une ville qui ne soit pas Paris?' he asked the industrialist:

> Il me répondit: 'Paris m'est nécessaire; c'est le réservoir humain qu'il me faut pour que, suivant les carnets de commandes de mes agents, j'embauche ou n'embouche pas!' Et un geste expressif illustrait ces paroles créatrices du bonheur ou de la misère des familles de ses ouvriers, selon la prospérité de l'usine. (2)

Businessmen and politicians recognized Paris's importance to the nation. Since industrial placements throughout the suburbs attracted the largest influx of new residents, that is where the next round of urban reform needed to happen.

As Guérard, Sarraut, and Chardon documented the rising influence of the suburbs, others feared the condition of Greater Paris's urban infrastructure, especially when compared to foreign rivals. In 1932, Louis Dausset, President of the Comité supérieur d'aménagement de la région parisienne, penned a letter to the journal *Urbanisme*, in which he lamented the difference between Paris and its German counterpart:

> Qand [*sic*] nous amenons des étrangers dans notre banlieue parisienne, ils sont stupéfaits: si à Berlin, dont on vous a parlé tout à l'heure, vous allez dans les quartiers excentriques, les quartiers des révolutionnaires, des communistes, dans les quartiers des miséreux – et Dieu sait s'il y en a! – à Berlin, on ne s'aperçoit de rien: on voit des rues, des voies aussi larges que les voies du centre; on voit des maisons qui sont un peu monotones, peut-être, mais il n'y a plus de ces rues tortueuses comme celles de notre vieux Paris, non il n'y a plus de taudis infâmes, mais des logements propres dans des maisons vastes, le long de spacieuses avenues bien alignées, qui ne donnent pas cet aspect de misère dont nous sommes témoins quand nous traversons certains de nos quartiers. ('A propos,' xix)

If unchecked, residents outside Paris may eventually push into the city centre, now occupied primarily by the bourgeoisie. Although the city's workshops needed labour, politicians hoped that a new master plan for the city would limit cross-border interaction and keep the 'other' out of the city. Dausset described that space:

> La banlieue, au lieu d'être l'exutoire, au lieu d'être, pour ainsi dire, le 'canal de dérivation,' que l'on empreinte quand on veut qu'un fleuve ne déborde pas, la banlieue devient pléthorique, devient surpeuplée, et déborde sur Paris. Et pour ne pas gêner des intérêts privés, pour que les gens qui travaillent aux Halles puissent coucher aux Halles, pour que les gens qui travaillent dans tel ou tel quartier puissent demeurer et coucher à côté de leur domicile, Paris s'engorge quotidiennement et le problème, à bref délai, va devenir insoluble. [...] Il s'agit donc à l'heure de dégager Paris, plutôt que de lui tracer un plan idéale, et M. Prost, qui travaille à ce plan, le sait mieux que nous! ('A propos,' xx)

In France, these two unequal but opposing sides competed for financial and political resources and suburban needs habitually came second. Although commentary found in urban journals tried to paint the broad suburban belt as uniform, each suburb had its own intrinsic characteristics. One town may have a factory that pollutes a local stream or possessed an infamous large shantytown. Another *commune* nearby, on the contrary, might contain small, well-kept homes. A few rural villages in the area even had adjacent large fields and meadows (Demangeon 47). Overall, the numerous *banlieues*, encompassing all economic milieus, remained disorganized and often antagonistic towards each other. This compositional structure made it difficult to collaborate on common interests. Beyond demography, technology also greatly changed the Paris-suburb dynamic. Just as railroads transformed one-time bucolic outposts such as Asnières into industrial towns, individually owned automobiles, products mainly for the nation's bourgeoisie, opened up much of the outer territory around Paris to new development, thus enhancing an already existing economic divide.

This dichotomy continued to be a source of consternation for many politicians. Albert Sarraut bristled when he thought of how dilapidated some suburbs remained:

> Quand j'ai entrepris de visiter, aux abords de Paris, ces lotissements misérables envers lesquels à la nation entière vient de faire un geste de solidarité, j'ai été frappé par le développement anarchique des cités suburbaines, par cette prolifération qu'on est malgré soi tenté de comparer à celle des cellules cancéreuses [...]. J'eus à cet instant le sentiment très net que les pouvoirs publics n'avaient plus le droit d'attendre, que toute temporisation nouvelle aggravait le mal. Il est superflu d'insister sur un phénomène que vous connaissez tous et sur les rapprochements pénibles qu'il suggère, lorsqu'on songe à la belle ordonnance de certaines banlieues étrangères. (*Compte rendu* 1)

The poor state of housing raised concerns. Henri Chardon warned that 'En Angleterre, en Allemagne, en Italie, en Belgique, presque partout les Gouvernements s'étaient rendu compte que les haines sociales naissent de la misère engendrée par les taudis bâtis au hasard d'infâmes lotissements' (1). The suburbs' increase in size and population throughout the Third Republic, along with political victory in 1924, meant that in all practical terms, government officials could no longer ignore them. Without more attention and reform, additional *ceinture rouge*-style political victories might occur and further disrupt the political status quo.

Beyond the mental or positional, Paris's defensive wall, the official physical city-suburb divide had kept out people from the 'cellules cancéreuses' lamented by Sarraut. France originally manned the 1840s barricade with guns directed out towards potential foreign adversaries. However, once technology rendered the walls outdated, unregulated development filled in the adjacent empty zones. Those weapons meant for the German army now faced suburban residents, which perfectly symbolized the antagonistic relationship between *intra-* and *extramuros* Paris. The eventual dismantling of Paris's fortifications throughout the 1920s began to erase some of the suburban marginalization and brought some communities into the region's urban conversation for the first time. Once France demolished the wall, the built-up physical definition of Paris could expand to incorporate both towns and land considered rural just decades earlier and create what would become known as 'Le Grand Paris.'

The term *banlieue* did not always equal squalor, misery, or overcrowding. Just across the English Channel, the idea of the suburb maintained a more positive connotation, although some religious leaders there argued against the creation of suburbs due to their 'modernist lifestyle.'[3] The evolution of the *cité-jardin* movement represented a counternarrative of suburban growth. Following the First World War, 'le goût pour l'habitat individuel commence à se répandre [en France], à l'imitation des pays anglo-saxons' (Bastié, *Urbanisation* 113). In the 1920s, since housing in central Paris was expensive and in short supply and automobiles were becoming more common for French families, a home of their own in the suburbs was becoming a positive, obtainable goal (Dennis 180). Most suburbs, however, continued the same economic segregation found inside the city. Wealthy villages, primarily to the city's west, contrasted greatly with the massive industrial outposts to the north and east: 'Les lieux d'implémentation bourgeoises en banlieue se distinguent assez nettement des concentrations ouvrières mais également, de façon

3 The Dean of St Paul's Cathedral condemned what he called 'bungaloid' growth. Artists and writers, including George Orwell, also lampooned suburban architecture and mores (Dennis 180).

plus dispersée: à Boulogne, Gentilly …' (Soulignac 178). If done correctly, a life in the French suburbs could be comfortable. Sizeable plots of land in wealthier *communes* accommodated detached homes and adjacent gardens provided more room and privacy than apartments in the city. The drive to suburban train stations and/or to central Paris also became much easier, although more individualized, with a car and paved roads.[4] In addition, access to the country or seaside was less of a hassle if one avoided the contested city centre. With these factors at play, French politicians, like their English and American counterparts, began to champion the idea that middle-class families too could occupy suburban land. Ideally, 'le centre de Paris serait alors réservé à ceux qui ont à faire dans Paris, durant la journée et qui peuvent ensuite s'en aller par des voies rapides dans la banlieue pour y passer des nuits fraiches et pour s'occuper de jardinage …' (Dausset, 'A propos' xx). With the city reserved for work and open space in the suburbs for residents and recreation, the new French suburban homeowner would benefit considerably from both fresh air and proximity to the embellished centre. Of course, if bourgeois arrivals altered the political make-up of the *ceinture rouge* to render it more favourable to centrist politicians, that would be an additional benefit.

As the suburbs began to shed some of their negative subtext, officials could see them as valuable partners in Paris's economic growth. Daily routines perhaps did most to change people's conceptions of city and suburb. The space created and identified by social practices such as 'everyday journeys to work, to home, to sites of leisure or consumption, the roads and the transport practices [soon dominated] material life' (Thacker 20). As the movement of people and goods between the city *intra-* and *extramuros* increased, it allowed both the familiarization and the formalization of the term and the concept of *la région parisienne* to gain ground. Greater exposure to the belt of development outside the capital – through newspaper stories, commerce, and personal experience – meant that the idea of a suburb, such as Chelles, and Paris forming one common urban entity steadily gained acceptance. Henceforth, successful planners would need to focus their attention on the previously ignored suburban ring where most new growth and

4 'Les classes moyennes [en France] peuvent déjà aller résider en banlieue, à condition de ne pas être à trop grande distance d'une gare' (Bastié, *Urbanisation* 114).

development would happen and incorporate movement between the two sides into their designs.

The Prime Minister's experience in Chelles in 1928 demonstrated the failure of previous Third Republic administrations 'to manage the conduct and welfare of [any] potentially insurrectionary urban populations by reorganizing and regulating [their] space' (Donald 51). However, by mapping the suburbs through plans like Prost's, governments hoped to understand the area's needs better and to avert any potential manifestations of working-class discontent. Put more simply, planning officials needed a capital city that could interact efficiently and knowledgeably with both the surrounding region and the nation.

Understanding the Terrain

In order for planners to control, know, and appreciate what was happening beyond the city limits, they needed to be familiar with the territory there. For Barthes, a city 'La cité est un discours, et ce discours est véritablement une langue: la ville parle à ses habitants, nous parlons notre ville, la ville où nous nous trouvons, simplement en l'habitant, en la parcourant, en la regardant.' (*Semiotics of the City* 92). Out and about in a city or neighbourhood, a pedestrian interacts with an area and this contact between inhabitant and city creates a dialogue. For much of the Belle Époque, Paris's urban conversation had been limited by the bourgeoisie primarily to exchange of ideas between the government and city centre. It was not until the twentieth century that planners like Jaussely even included the *banlieues* in their comprehensive projects. Before these plans, few saw the need to acknowledge existing or potential development beyond Thiers's fortifications. This lacuna reduced or eliminated the suburban ring's presence in people's minds as well as in the larger urban debate and meant that the area received less cartographic representation. With the exception of the military's *état-major* maps, which the army originally produced for its own needs, most official maps of Paris in the nineteenth and early twentieth century dealt almost exclusively with the city alone. Plans, such as those published by Hénard or Alphand,[5] presented the city

5 Alphand's *Paris en 1889* atlas is discussed in Chapter 3. Hénard's *Études sur la transformation de Paris* (1904) is described in Chapter 5.

up to its border and often left any remaining map space blank. This separation between official planners and suburban inhabitants left a sizeable proportion of the region's population and territory – that same area highlighted by Guérard back in 1925 – isolated and ignored by anyone of influence. By 1934, the suburbs had surpassed Paris and received more of the government's consideration.

Micromanaging a city or urban space can result in the creation of an over-stylized, banal prestige district with little practicality for its populace. Planners who operate in this fashion often 'misrecognize not just the social relations, but the very space of the city. They see it as a territory to be bounded, mapped, occupied, and exploited, a population to be managed and perfected' (Donald 121). Heavily embellished central Paris, with its monuments, cafés, and department stores, best symbolizes this type of over-built urban area. Such a district attracted domestic and foreign visitors but also monopolized the city's attention and finances.

Impressive as central Paris was, one can also argue that by 1934 it had also lost some of its usability. Large, imposing monuments served little purpose to morning commuters and contributed modestly, if at all, to the daily operations of the city. Instead, the Republic appropriated and used this area as a stage to highlight its achievements, even if that meant less time, attention, and money to better the suburbs, where most of the population growth occurred. Large spectacles such as the 1925 Exposition internationale des Arts décoratifs et industriels in the Champ-de-Mars and Trocadero and the 1931 Exhibition coloniale internationale at the Porte Dorée and the Bois de Vincennes highlighted the capital, its open spaces, and its ability to host millions of visitors, but did not improve traffic congestion or the housing situation in the suburbs.

Unregulated suburbs contrasted greatly with that embellished city centre and few tourists ventured there, assuming they could even find their way around them since, historically, cartographers and publishers had preferred to map away problematic *quartiers*. Once map-makers silence or eliminate troublesome neighbourhoods or communities, that threat is gone, on paper at least. However, as the 1871 Commune de Paris demonstrated, large industrial French cities remained unpredictable and often too complicated to corral, manhandle, or simply map into submission. Just because a shantytown failed to appear on a map does not mean it did not exist or affect surrounding neighbourhoods. So, attitudes had to change. Instead of ignoring towns or villages, the government needed to acknowledge the benefits of documenting these once-forgotten areas previously silenced by city planners. Even as city councils began to survey their problematic neighbourhoods, government planners operated from offices physically separate from the quarters or suburbs they hoped to transform. This mental and physical distance allowed them to make unpopular or impractical decisions with little or no accountability to the residents affected.

Upon passage of the 1919 loi Cornudet, regulation and planning moved from under local edicts to national directives, but initially little was done in communities and villages throughout the Seine and Seine-et-Oise departments, where it was needed most. Paris itself never implemented the Jaussely plan, but other larger French cities did do what the legislation had required of them. Of the 1,600 French *communes* to have published plans by 1931, 250 of them (about 15%) submitted their proposals to the Ministry of the Interior for approval. In 1910 Lyon published its project, well before it was legally required to do so. In 1919, Reims, which had suffered considerable damage in the war, was one the first to have its project adopted by the Commission supérieure des plans d'aménagement in accordance with the loi Cornudet. Lille followed in 1920. Two years later, Dunquerque and Briançon published their plans. In 1923, maps from Toulon and Aix-les-Thermes were submitted. Of the 138 *communes* in the *région parisienne* to have drawn up plans, only 75 eventually received approval to implement them (Delorme 28). Since so few towns and cities complying with the law, the region needed comprehensive development.

Legislating Change

The patchwork of ineffective stop-gap laws during the 1920s hindered comprehensive regional Parisian planning (Bastié 116).[6] Prime Minster Poincaré's experience of the *banlieue* in 1928 restarted the debate, and the discussion on how the government could most effectively tame and manage the 'le Grand Paris,' primarily its suburbs, resumed.

6 Gaston Monsarrat provides a list of some of these measures in his explanation of the Commission supérieur d'aménagement, d'embellissement et d'extension des villes in the March 1932 special issue of *Urbanisme* (xiv).

Later that same year, at Poincaré's suggestion, a Comité supérieur de l'aménagement et de l'organisation générale de la région parisienne (CSAOG) was formed. Albert Sarraut, then Minister of the Interior, hoped that the new committee would address underdevelopment and cure what he referred to as the suburban 'sickness,' poverty. Most important, even though the Service géographique had surveyed them in the first few years of the century, officials at all levels would now get an up-to-date account of the suburbs (Remaury 148). Sarraut asked the committee members to organize the *région parisienne* and create 'un régime d'exception. Il faut donc tout d'abord la définir, tracer autour de Paris une ligne à l'intérieur de laquelle certaines choses, permises ailleurs, seront interdites' (4). The creation of a special Paris zone would set the capital apart from other cities and identify and solve its problems. It appeared that Sarraut was initially more concerned with regulating factory construction in the region than anything else. The first 'interdictions' he proposed were, indeed, 'des usines, qui sont l'une des principales causes de la surpopulation.' In addition, the government needed more 'réglementation des établissements incommodes et insalubres' (4–5). Dirty factories not only attracted too many newcomers, they also produced pollution and worsened living conditions for neighbouring residents and the government needed to control and regulate them better.

With the Seine department as well as parts of the Seine-et-Oise and the Seine-et-Marne considered part of the 35-kilometre 'Le Grand Paris' by the 1932 law, interdepartmental cooperation regarding industry and housing became easier. The CSAOG itself had been given a triple mission: 'Sa mission est donc en premier lieu une mission d'investigation et de contrôle. Il se saisira des projets dont la réalisation est de nature à modifier l'organisation ou l'aspect de la région parisienne. Il en signalera les inconvénients éventuels, il suggéra les modifications souhaitables' (*Compte rendu* 4). Sarraut acknowledged that previous legislation, such as the 1919 loi Cornudet, failed to plan French suburban growth. He called the CSAOG to task. The committee's role, 'va être de mettre de l'ordre dans l'immense agglomération parisienne, de prévoir son extension ultérieure. Il y a ici, tout d'abord, un fait dont nous devrons nous préoccuper, l'afflux de la population' (*Compte rendu* 2). He started by dividing the committee into five sections to best develop urban planning legislation for the capital. The first group (Aménagement) gathered all plans and ideas; the second (Législation) took those submissions and prepared legislation that would apply specifically to the Paris region; the third section (Finances) looked for available resources and grants as well as examining the financial implications of the potential projects; the fourth faction (Services publics) coordinated public services and transportation; and the fifth and final committee (Habitation) investigated the region's acute housing shortage (*Compte rendu* 7–8). Henri Prost himself was part of the first group and 'son expérience du Var l'avait familiarisé avec l'aménagement des groupements de communes solidaires, les réservations de terrains, la sauvegarde des sites, la lutte contre l'envahissement de certains territoires. Il apportait aussi l'extraordinaire sens des "réalités" que lui avaient valu ses dix années d'urbanisme appliqué vécues au Maroc' (Remaury 150). Due to his experience, his vision for Paris received much support from fellow CSAOG members.

Citing the need for reform, Gilles Monsarrat, director of the Ministry of Interior, listed five categories of Projets d'aménagement et d'extension on which the committee would comment. The most important, the fifth, concerned the capital:

> V: *Projets d'aménagement des communes situées dans la région parisienne ou dans une autre région devant faire l'objet d'un plan régional.* – le projet de loi déposé par le Gouvernement le 12 juillet 1929 et qui tend à autoriser l'établissement d'un projet d'aménagement de la Région Parisienne dispose dans son article 7 et 5: 'Les projets d'aménagement communaux pour lesquels la déclaration d'utilité publique est intervenue antérieurement à la date où sera déclaré d'utilité publique le projet régional ou le projet partiel seront modifié, s'il y a lieu, sur l'initiative du Préfet ou du Maire et mis en concordance avec le projet régional.' (xvii)

Monsarrat's declaration, conferred by Dausset and seconded by M. Camille Rosier, the committee's secretary general, explicitly lists the goals of drafting the 1932 legislation and cautions readers that the new law adheres to the former governmental regulations of 1929 and, if city plans did not conform, city halls would be able to modify previously submitted projects.

Six years after Poincaré's troublesome return to Paris, the National Assembly passed la loi du 14 mars 1932. The broad planning legislation officially defined 'la région parisienne' geographically for the first time, not demographically, as the area within a 35-kilometre radius from the *parvis* of Notre-Dame (Lavedan *Nouvelle histoire*, 155). In the legislation, the government established the Comité supérieur d'aménagement de la région parisienne (CSARP), previously known as CSAORP. The law contained many articles, of which the most pertinent for this project were:

1 Le projet régional d'aménagement de la région parisienne sera soumis à une enquête qui s'ouvrira sur le plan directeur et sur le programme établis par le comité supérieur de l'aménagement et de l'organisation générale de la région parisienne et provisoirement arrêtés par le ministre de l'Intérieur.
2 Cette enquête aura lieu d'abord dans chacune des communes qui sont comprises dans la région parisienne.
3 Pour chacune des communes auxquelles s'applique l'article ci-dessus du précédent décret, il sera constitué par les soins du préfet un dossier comprenant les pièces suivantes:
   a. Un exemplaire ou une reproduction réduite du plan directeur de la région parisienne au 1/50.000;
   b. Un extrait au 1/10.000 ou au 1/20.000 de la partie du plan directeur concernant la commune intéressée et les communes limitrophes;
   c. Un exemplaire du programme des servitudes prévu par l'article 2 e la loi du 14 mai 1932 avec ses annexes;
   d. Un mémoire descriptif faisant apparaître le caractère et les grandes lignes du projet régional d'aménagement.

The most important component in terms of cartography was the fact that each affected community needed to complete a comprehensive report, in which officials should find at least one map of the Paris region at a scale of 1/50,000. This large-scale depiction of Paris and the suburban surroundings was to be accompanied by smaller, more detailed maps as well as a description of the overall project and the transportation axes.

Once the details of the law had been released, local elected officials throughout the suburbs voiced their opinions and joined the discussion. André Morizet, the mayor of Boulogne-sur-Seine, for one, hoped the guidelines would produce a new Haussmann-like personality and bring drastic change to the region. However, he realized that the current political climate ensured that one powerful voice alone would not be able to dictate a radical new plan. Instead, he anticipated a collaborative final product: 'Lorsqu'on saura que les élus de la région parisienne n'exposent pas seulement des idées justes, mais qu'ils expriment dans les questions d'aménagement la volonté formelle de six à sept millions de citoyens, les difficultés de tous genres pèseront peu' (400). Looking out for the region's six to seven million residents would best ensure that the regional plan benefited the largest number possible.

Louis Dausset reinforced this idea, stating that 'Il est absolument impossible en effet, de faire de l'urbanisme si on n'a pas d'idée, la notion de la plus value qu'on donne au terrain, et de la contribution que les bénéficiaires de cette plus-value, souvent excessive, doivent apporter aux œuvres d'intérêt public' (xix). Burned by large-scale public works programmes in the past, such as Haussmann's, which seemed to benefit a small percentage of the population, Dausset wished to make sure that the general public, not leaders or businessmen, would gain most from the new legislation.

Following the urbanism regulations in vigour since passage of the 1919 loi Cornudet, such as opening up urban plans to competition and incorporating ideas discussed for 20 years by the Société française d'urbanisme (SFU), the 40-member commission charged with drafting a comprehensive plan and presided by Louis Dausset, looked over and chose Prost along with Pierre Remaury and Jean Royer to guide the region forward. The 1932 law also stipulated that the project 'ordonner et vertébrer' the Paris metropolitan area (Laisney 119). The idea that one person or project would finally impose order onto the chaotic suburbs was both important and welcomed. A seminal document for the Paris region, it would change the face of Paris urbanism.

Two weeks after the passage of the 1932 law, the Comité supérieur assigned Prost the full title of 'Urbaniste en chef responsible' and he began working on his plan. The law mandated a two-year deadline for the final draft to be presented to the government. Twenty-six months later, on 14 May 1934, Henri Prost submitted his Plan Prost to the Minister of the Interior. As a result of Prost's study of the area, new concepts appeared on a Paris urban plan. The first, the adoption of a larger metropolitan area border had tremendous significance. The official 35-kilometre perimeter around Greater Paris would provide at least three short-term benefits to the region: it would harmonize zoning to stabilize real estate values, establish communal financing for large-scale projects, and protect rural and wooded areas from unregulated development. A second notion, the creation of a network of limited-access roads, suggested by Jaussely back in 1919, would help improve the movement of people and goods in and out of the Paris region and avoid the problems experienced by Poincaré back in 1928 as he returned from the countryside (Royer, *Région* 23). For the first time, Prost provided guidance for building outside the city limits and explained how improvements there would benefit all residents. Second, incorporating the new territory in his official map altered how the public would use and view the whole region, not just the city itself. The third, and perhaps the most important, component of this plan was the fact that the government expressed, in writing, the explicit goal of understanding and knowing the entire Paris territory, an aspect lacking in the 1919 legislation.

The truly groundbreaking nature of Prost's project and its contents did not mean that all politicians agreed with the ideas and the plan of action found inside. Dausset foretold the problematic fate of the plan:

> Nous avons, au Ministère de l'Intérieur, au Comité Supérieur de la Région Parisienne une œuvre très ingrate, je vous assure; le plan d'aménagement parisien de ce qu'on appelle le Grand Paris, a été l'objet d'un projet de loi que nous avons élaboré pendant plus de deux ans, qui finit par être déposé en 1929. La Chambre a mis un an à le voter, encore avec des amendements un peu dangereux; ce projet de loi a été discuté au Sénat. La Commission des Finances n'a pas voulu accorder l'autre jour les quelques millions nécessaires à exécuter le plan de M. Prost, c'est-à-dire à transformer le plan de base en un plan précis. (xx)

Backlash against the government's large-scale project started soon after Prost released his *Carte générale*. The fact that the Commission des Finances did not wish to fund the plan was the first impediment of many. Yet there were positive aspects to Prost's ideas.

Not only did Prost provide the Conseil supérieur with detailed knowledge of the Seine department along with various *communes limitrophes*, but he also certified the need to control the region's growth. Upon publication and approval of this project, the age of unlimited French urban sprawl officially ended, at least on paper. The 'urbanism' in Prost's plan constituted an important component in the birth of planning and no longer one of infinite extension: 'L'urbanisme des Congrès Internationaux d'Architecture Moderne (CIAM) opposé à cet urbanisme néo-haussmannien refuse délibérément cette notion d'aménagement au profit d'une vision radicale de la rénovation des centres historiques et de l'extension indéfinie de la périphérie' (Delorme 27). From this point forward, Greater Paris urbanism would not be limited to the restoration and modification of its historic centre accompanied by uncontrolled suburban expansion. Instead, planners would order and manage the extension of roads and infrastructure out from the city.

After agreeing with his viewpoint, the French government accepted Prost's map, which coordinated regional planning for its capital city. Paris and surrounding *communes* were now bundled into one cohesive administrative unit, better matching the area's demographic reality. Perhaps more significant, a new mental border, 35 kilometres from Notre-Dame Cathedral, had now been designated, replacing the former physical wall adjacent to the city and all-but-absent cartographically in 1906 and 1919. This new geographic legal entity would facilitate large-scale infrastructural improvements and prohibit individual *communes* from disrupting projects, such as an *autoroute*, that might benefit the region as a whole. Parisian administrators would look forward to create an effective and realistic planning proposal that would foster organized development and growth.

Prost's Cartographic Discourse

When Prost presented his plan to government officials, he assured them that the project's goal was not just to facilitate urban growth. He did not provide Paris with a *plan d'extension*, as Jaussely had in 1919. He offered instead a 'véritable plan de réaménagement' with the aim to enhance access to Paris, address traffic congestion via new limited-access roads, and assure the preservation or wooded areas and open spaces (*Le Petit Parisien* 1934). His *Carte générale* presented these goals and showed officials how he would change and improve the region. One of the first items the reader sees on Plan I (Région nord) is the project legend, which identifies how Prost presents his ideas to officials. Prost needed the map to demonstrate not only how he viewed the metropolitan area and its future, but also to do so in a coherent manner that readers could reasonably follow. Limited by available map space, the legend organizes and labels Prost's cartography. It also explains to readers how to navigate the document. Prost's legend contains ten categories and a scale and clarifies his vision. It establishes and then prioritizes map components, but in a different order from the Service géographique, the other legend discussed in this book. Most important, via text and legend, Prost proposed solutions for three of Paris's major problems: the lack of quality housing, protecting green space, and the need to facilitate movement in the region.

His plan documents existing residential and industrial development, parks and forests, and roads, as well as intentions for future expansion. The first bar, in orange, represents 'Habitations.' Following convention, as the item encountered before all others, housing infrastructure gets the legend's 'best' spot. In placing it there, Prost acknowledged the administration's priority of documenting the growth of the unregulated suburban housing stock. Furthermore, Prime Minister Poincaré's experience demonstrated that laws throughout the 1920s had failed to address the region's lack of decent accommodation. He assigned the second bar, 'Usines, terrains industriels, voies ferrées,' a dark-red colour, reminiscent of the Service géographique's 1906 *feuille* XXIII_14. Prost, some may argue, assigned the region's industrial infrastructure a political hue. The *ceinture rouge* represented what officials feared most: a large, concentrated working-class population, one that, if provoked, could upset political stability. This grouping is second on the list, giving it prominence over the remaining categories, almost as if Prost was reminding readers of its menacing nature. Furthermore, with this tint, Paris's industry stands out on the map. Any official with access to the *Carte générale* could identify industrial clusters and know where to locate possible trouble spots. Although the government feared the proletariat, French industry still provided many jobs at a time when both were in short supply.[7] By 1937, 'la forteresse,' the Renault factory in Boulogne-Billancourt, one such source of potential trouble, employed more than 37,000 salaried and 30,000 factory workers. The complex housed large numbers of personnel and was also a hotbed for organized labour, which 'devient une sorte de baromètre qui renseigne sur les réactions de toute la classe ouvrière' (Soulignac 83). And the threat was real. In 1936, just two years after Prost's map was published, over a million French workers went on strike, shutting down vital

7 The 1931 economic crisis greatly affected suburban industry. Unemployment reached between 15 and 16 per cent of the population, more than 200,000 people (Soulignac 85).

industries and winning important labour concessions in the Matignon Agreements. Many politicians, wary of the increased organization and militancy of France's working-class population, had to look back only to 1871 to see what could happen if too many unhappy residents began to voice their discontent, and feared that such an uprising was a possibility. Nevertheless, one key component of the new legislation was the aim to better 'know' all aspects of the region. Prost could not ignore a significant portion of 'le grand Paris' and his map delivers the mandate with detailed information.

Prost's legend next displays green 'Forêts, bois, jardins publics,' which affirm that the ever-developing region still possessed a significant inventory of parks and forests regardless of the extent of new development and growth. In the early twentieth century, large swathes of forests – such as Bondy, visible as recently as Jaussely's 1919 *Plan d'extension*, 15 kilometres to the east of Paris – disappeared and were replaced with rows of houses (Royer, *Région* 21). At the same time, officials recognized the importance of non-developed land for the region's population in the crowded urban zone. The Musée social's championing of green space for the health and well-being of the area's residents meant that Prost needed to find a way to incorporate existing (or even find new) open space in Greater Paris as it ameliorated residents' health, their quality of life, and the area's aesthetic. In an era where terms like 'fortress' were used to describe large factories, open land made for a nice buffer between order and chaos. Guérard, for one, believed that green should combat red: 'Le remède contre "la ceinture rouge," ce n'est pas la répression, ni même la restriction, mais le mieux-être [...]. Tout cela peut se traduire ainsi en termes d'urbanisme: faites de votre ceinture rouge une ceinture verte' (169). For Guérard, parks and trees should replace Thiers's fortifications and, as a bonus, buffer the city from undesirable populations.

To make these prized spaces highly visible on his map, Prost employed a conspicuous tint whereby leafy woods and parks signal a calm escape from the dust and noise of urban development and his green bar contrasts greatly with the city's built-up terrain. The associative nature of the colour, used for trees and grass in the maps examined in this book, represents a cartographic norm in Third Republic cities and offers a marked difference from the red factories and orange housing. Furthermore, the crowded industrial suburbs too wished to have better access to parks. For those living in insalubrious housing, a nearby wooded area or park might be the only respite from dark, dirty living conditions. So, the green areas on Prost's map appealed to everyone.

The next category of land in the legend is 'Terres, terrains non bâtis' and the yellow bar symbolizes the undeveloped land that its name suggests. This inclusion marks a significant change from 1889, 1906, and 1919, when cartographers often ignored undeveloped property. However, by 1934 Prost demonstrated the need to identify vacant land. With the region's population increasing each year, both developers and government officials had to find a way to accommodate new homes and businesses along with accompanying infrastructure.[8] The accurate location of available undeveloped territory would help the Comité supérieur easily identify where to direct new construction or in which zones less-desirable projects, such as a noisy train yard, could be placed. It would also help them avoid the loss of more green space. For an urban area constantly absorbing newcomers, the perceived value of undeveloped land had changed since Alphand, the army, and Jaussely published their maps. On their plans, Alphand and the military discounted undeveloped land cartographically. Its non-value was reflected in the lack of colour. Jaussely, on the other hand, covered much of his map with paint. To him, open land meant that the government had nothing to order or manage and did not merit his attention.

The final two bars in Prost's legend, the 'Emplacements destinés à des promenades publiques ou terrains de jeux' and the 'Emplacements destinés à des promenades publiques ou terrains de jeux, soit à des services publiques, soit à des cités jardins,' represent not 1934 Paris and its environs, but the future. To differentiate between the two, Prost modifies his standard green portrayal and adds crossed lines, an act that will allow the reader to distinguish between existing and potential free space in the ever-crowded region and show officials and readers both where and how Prost would improve access to green space.

His right-hand column, in its slightly less important location, contains three components. A red line indicating 'Routes automobiles

8 Between 1880 and 1914, the western and north-western suburbs, from Boulogne to Saint-Denis and from Gennevilliers to Asnières and Columbes, experienced the greatest development. There, 'le relief n'offre aucun obstacle et sur ces terrains plats les voies partant des gares Saint-Lazare et du Nord rayonnent largement en pattes d'oie' (Bastié, *Urbanisation* 113).

à circulation réglementée' occupies the first, and therefore most important, spot. Following these limited-access roads, Prost lists 'Routes principales' with two bars shaded orange. Finally, a red circle demonstrates where the reader would encounter 'Points de raccordement et de jonction où les courants de circulation pourront être éventuellement séparés.' These bull's eyes promise the reader dedicated Paris entry and exit points between different tiered roads, which should enhance traffic flow considerably.

Prost finished with an 'N.B.,' with no corresponding sign or colour. In it, he advised the reader that he had not incorporated everything into the legend. He stated that 'une voie projetée est représentée par un trait de couleur bordé d'un liseré noir continue, une voie existante à élargir, par un trait de couleur bordé d'un liseré noir ponctué, un ouvrage important par un trait de couleur hachure noir.' Contrary to his green park 'emplacements destinés,' he did not feel the need to identify planned road construction in the legend. Instead, he indicated future route changes only on the map itself. His employment of narrative to describe a non-mapped component, however, breaks cartographic norms. Prost asked his readers to turn away from his colours and symbols with fixed definitions, read this text, return to his map, and apply their existing cartographic knowledge to identify black marks next to routes and highways as future road construction.

Of course, a request such as this was not completely unknown. The Service géographique had used the same technique to instruct its readers about Parisian topography on the 1906 *Carte de France*. It affirms a comfortable map-reading ability among the Comité supérieur members. However, the goal of a cartographer should be to make map reading as easy as possible. Each additional step, such as this, increases the chances of error. For those readers who ignore or miss his statement, the regional road network appears as a *fait accompli*, even if much of it remained to be built. This break from standards may also affirm Prost's confidence in the eventual construction of these roads as a cornerstone of the new legislation.

Roads

Prost's legend affirms his cartographic hierarchy in the placement of 'Habitations' before all other categories. Yet his Plan II colour use affirms that Prost wished to enhance access to the capital via large limited-access roads.[9] As one would expect, grey rail lines can be found throughout the region. However, by 1934, the automobile assumes command of his narrative. The stark dark-red 'Routes automobiles à circulation réglementée,' although skeletal in nature, along with his 'Points de raccordement et de junction,' are the most prominent map elements, affording them considerable significance. The 1930s route infrastructure contrasts greatly with the muted yellow, orange, and red land colours. On paper, this network provides easy and abundant access to the city. Yet, on the ground, Parisian area drivers experienced the contrary. In his letter to the journal *Urbanisme*, Louis Dausset listed traffic congestion as one of the main problems when entering or leaving the capital: 'Paris ressemble, je l'ai dit bien des fois, à une bouteille. Quand vous sortez par la route d'Italie vous entrez dans le fond de la bouteille, puis vous arrivez à Villejuif, et, là … vous tombez dans le goulot!' (xx). In his regional plan, Prost addressed this concern. He envisioned the construction of five *autoroutes* to complement France's existing infrastructure. When built, these limited access highways would help drivers avoid the crowded suburbs along with numerous small villages and terminate at the city's edge.

What does his road structure suggest about 1930s Paris? Haussmann's nineteenth-century Parisian renovations embellished the city centre and limited bourgeois suburban flight in much of the early Third Republic. By the third decade of the twentieth century, the automobile allowed many French families to leave Paris in the search for more space. Tramways and trains moved many people from neighbourhood to neighbourhood in the second half of the nineteenth century. After the First World War, they began to move people from suburban train station to urban centre. Albert Demangeon, a member of the Commission administrative de l'Institut d'histoire wrote of this new Parisian reality in *Paris, la Ville et sa Banlieue* (1933). Based on transportation access, for him the Paris agglomeration covered the entire Seine department, as well as surrounding *communes* in the Seine-et-Oise and Seine-et-Marne departments:

9 This middle section of the *Plan du projet directeur approuvé* might be considered the most 'important' as it is the only one of the three to include the city of Paris itself.

> Dans son développement, [cette agglomération] s'appuie essentiellement sur les voies de communications. Avec la progression des chemins de fers et des tramways, les Parisiens ont su se répandre sur un périmètre de plus en plus étendu. La limite de leur résidence permanente n'est autre que la limite de la circulation facile et rapide, déterminée elle-même par la densité du réseau des transports en commun, par le temps du parcours et par les frais de voyage. Chaque tentacule que la grande ville pousse vers l'extérieur correspond à une ligne de communication. (Demangeon, qtd in Bellanger 217)

In conjunction with mass transit, the personal automobile had a significant effect on the changing suburban composition as it afforded car owners a newly found sense of freedom: the ability to commute independently from set rail routes and schedules. In 1919, Jaussely placed roads next to rail as it helped forecast where population growth would take place. By 1934, roads broke free from the train and instead followed the bourgeoisie as it spread further throughout the region. In turn, these newly constructed highways facilitated more residential growth. The wide breadth and range of the 'routes automobiles à circulation règlementée' on Prost's map confirm the need and desire to access all corners of the region.

Of course, not all Parisians looked to escape the city. Paris's central and western *arrondissements* retained much of their prestige and continued to house significant bourgeois populations, especially with the large numbers of highly desirable 'beaux immeubles que les sociétés immobilières construisent' near the edges of the Bois de Boulogne and on the Champs-Élysées (Bastié, *Urbanisation* 117). Paris remained the home to the region's upper-middle class, perhaps suggesting why Prost preferred to keep the red limited-access roads and their potentially disruptive construction almost exclusively outside the city. Eugène Hénard, the last individual to propose a large-scale renovation of the city centre, had radical ideas, but their potential cost and disruption meant that the city never seriously considered them.[10] To facilitate automobile access to the city, Prost injected two main roadways into Paris. One entered the city limits from the north-west. He divided it

10 For a discussion of Hénard's view for Paris, see Olson, 'A New Paris' and Wolf.

into two small arrows pointing directly at the city centre via the Gare Saint-Lazare, an area which Raoul Dautry, as director of the Chemins de fer de l'État, modified and Prost knew well (Royer 21). The second roadway barely punctures the city's southern periphery, where a third arrow directs readers up the boulevard Raspail towards Montparnasse train station. Here, Prost further subtly confirmed the continued importance of railways, a somewhat overshadowed component on this project's map, to the city and region as he employed red to highlight two prominent Parisian entry points used by suburban commuters.

His other main arteries skirt around and avoid Paris. To the west of the capital, adjacent to both the Bois de Boulogne and Paris's upscale sixteenth *arrondissement*, Prost charted the *autoroute de l'ouest*. When complete, that limited-access road would take drivers out of the city, past the communities of Versailles and Saint-Germain-en-Laye, two former royal residences. It provided the wealthier western half of the metropolitan area with enhanced access to the Normandy seaside resort of Deauville, the industrial port city of Cherbourg, and the regional capital of Rouen. A southern branch of that same roadway directs residents to Brittany (Rennes, Brest) and western Normandy (Granville, Le Mont Saint-Michel). Both routes serve popular summer *stations balnéaires*. The Western Autoroute, the first to be designed and built, starts just past Boulogne near the RN 307 and continues to Rocquencourt, where Prost divided it into two branches. One branch met up with the N190 near Chambourcy, the other joined up with the N10 near the palace gardens of Versailles. With these new roadways, it would be easy for bourgeois Parisian car owners to navigate through the inner industrial suburbs and quit the city with little or no exposure to them.

Another road addition was the extension of Paris's traditional *voie triomphale*, which Prost outlines in deep red. A project originally envisioned by Jean-Baptiste Colbert under Louis XIV and proposed again in 1919 by Jaussely, Prost too argued for the implementation of Paris's *grand axe monumental*, which starts at the Louvre and passes up the Champs-Élysées to the Arc de Triomphe. Here, he followed Jaussely's lead and proposed adding a 'route automobile à circulation réglementée' past the Place de l'Etoile, to approach the city's edge and extend out to the forest of Saint-Germain. However, Prost suggested that only the section between Saint-Germain and Nanterre would actually be considered *autoroute* grade. To complement its stature and location, Prost envisioned the grand avenue housing ministries, embassies and government buildings to become a 'Washington [D.C.] de la France' (Remaury 167–68). Prost recognized the importance of such a district to the city and region. Also outlined on his map were the *autoroutes du nord* and *de l'est*, both of which will traverse the same unregulated development experienced by Poincaré. The *autoroute du nord* would take drivers towards Lille, then onto Flanders and Holland. The *autoroute de l'est* would link the capital to Alsace-Lorraine and eventually Germany.

Paris's secondary road network, represented in orange, is less pronounced, but considerably more comprehensive. These *routes principales* support the region's automobile transportation infrastructure. Without them, the area's vehicular traffic, comprised of both individually owned cars and large commercial trucks, would be forced onto over-burdened surface streets, hindering potential growth and creating nightmare traffic jams. Regional highways and two beltways link the autoroute termini and connect Paris with the rest of the country. Diverse touristic routes were also considered and eventually included in the project (Remaury 174). In some respects, the orange roads, overshadowed by Prost's dark-red *autoroutes*, serve as the workhorse of his new Paris region. They blend harmoniously with the existing terrain and reassure any hesitant administrator worried about possible disruptions from construction, especially as 'les migrations journalières ne cessent d'augmenter' (Soulignac 85).[11] Since most of the planned *autoroutes* were designed to complement and join up with existing roads, many of the more central *routes principales* adjacent to Paris already exist. Prost included their *route nationale* number designations, and their incorporation into the enhanced road network is portrayed here as a *fait accompli*. Readers should already know many of them. Prost felt little need to spend any more of his cartographic capital on them.

The space that previously hosted the city's walled fortifications and which Jaussely slated for parks now includes an orange loop. Although

11 Soulignac states that 'un des phénomènes marquants de cette période et qui caractérise fortement la vie en banlieue, est, sans doute, la dissociation grandissante entre l'habitat et l'emploi. Dans le sens banlieue-Paris, [des trajets entre les deux] one été multipliés par 5 en 30 ans, atteignant 450 000 personnes en 1931. Dans le sens Paris-banlieue, [ils] ont également augmenté de 5 fois, mais à partir de chiffres plus faibles (70 000 personnes en 1931)' (85).

not yet constructed, Prost wished to create what would become Paris's principal *boulevard périphérique* in that zone. The proposed centrepiece of the entire regional road network where Thiers's former barrier once stood, this mirrored a 1904 proposal made by Hénard.[12] When built, it would allow automobilists to avoid the crowded centre and continue out towards the suburbs and provinces on any of the new *autoroutes*. Construction of the present-day *boulevard périphérique* did not begin until 1956, well after this map's conception, yet its presence here demonstrates the practicality and forward thinking of Prost and his plan. Furthermore, the inclusion of the boulevard in the 1934 map allowed Prost to minimize and replace the *ceinture* circular railroad that ended its passenger service that same year.

Prost had many additional ideas for the region's future road network and appeared prescient as he suggested much of Paris's current transportation infrastructure before it became reality. Indeed, the *boulevard périphérique* is not the only existing major road that Prost visualized. A reader can spot what would become the current *autoroute* A86, also known informally as the *super périphérique* as it surrounds both the city of Paris and its *petite couronne*, the first ring of inner suburbs on the *Carte générale*. The urbanist recommended the construction of Paris's second ring road some 30 years before its eventual creation, though Prost gave it a much wider trajectory around Paris than it occupies today. His *autoroutes de l'ouest* and *du sud*, depicted in both his supplementary sketches and on his map, would not lead Parisian motorists away from the crowded city to western and southern France until the 1960s, following the construction of Orly Airport. In addition to his comprehensive *réseaux routier*, Prost also revives the French tradition of majestic, wooded, tree-lined avenues, such as the Route de Versailles and Boulevard Impératrice, both of which joined Paris with communities further west via his large-scale, car-orientated, limited-access highways. Finally, Prost's 'Points de raccordement et de jonction,' various interchange circles primarily in the outer ring and other sectors of highways, link the two levels of road works. Once implemented, his Paris region would enjoy seamless compatibility between the major thoroughfares and the smaller *route nationales*. Prost's

12 Hénard's proposed a ring of *boulevards des fortifications* to encircle the capital. The map with this road can be found on page 211 of his 'Transformations.' (Hénard, 'Transformations' 211).

symbol for 'Points de raccordement' mimic traffic circles; a relatively new infrastructural improvement used for road interchanges and the dark-red circles further document Prost's desire to look forward.[13]

As car owners navigate the various *routes* throughout the metropolitan area, it appears as if all of Prost's roads lead drivers away from the city centre. To emphasize this ease in exiting the crowded city, he employed large white arrows to direct drivers away from Paris and the suburban chaos, and guide them to specific named cities, such as Rouen, Dieppe, Calais, Dijon, etc., whose font is large enough to be legible from a distance. Prost related the capital with regional cities. Any non-Parisian reader could find the fastest and easiest route home. With the arrows, Prost promised unrestricted interaction with the entire country and highlighted the Comité supérieur's original mission of defining Paris by 'grands courants généraux de circulation' (Remaury 152). Tasked to 'vertébrer la région,' Prost highlighted roads, but also emphasized movement. For a reader who experienced the same frustrating congestion described by Poincaré, the legislative goal is not only worthwhile, but sorely needed. Back in 1919, Jaussely first acknowledged the personal automobile's importance as he planned the future of the region. Prost now confirmed its position in everyday life and endorsed the capital's role as the hub of France's road network.

The Suburban Footprint

Once the Comité supérieur initiated Prost's plan, the suburbs would acquire significantly improved road coverage. Even though Jaussely proposed to develop the region in 1919, the government still needed to better understand the territory beyond Paris's city limits. Mapping the suburbs afforded the government two advantages. First, once the survey for buildings and roads had been completed, much of region's remaining 'unknown' would all but disappear. Second, when the planning agency documented the *ceinture rouge*, like Haussmann's triangulation of Paris before his renovations, its population could be monitored better and perhaps even controlled more easily. Although the 1871 Commune

13 One first such suggestion for Paris was Eugène Hénard's *Etudes sur la transformation de Paris*, in which he proposed multilevel traffic circles at major intersections.

de Paris rebellion grew partly from political discontent with the conservative Government of National Defence, it did erupt from economically depressed neighbourhoods. The national government knew that to prevent a similar manifestation from reoccurring in the suburban ring, it needed to keep this disadvantaged population from becoming an active opposition. This task would require knowledgeable and effective supervision of those residents, especially after the political victories of 1924 that handed them a larger voice in parliament. Prost needed to provide city and departmental administrators with at least the illusion of a governable space. His suburban ideal – roads and mapped land – offered a well-documented controllable territory.

Although traffic remained the Comité supérieur's most pressing concern, as demonstrated by Prost's bright roadways, the availability of housing in the region also continued to disappoint and was a factor in the government's wish to regulate land in the suburbs. To address this concern, Prost documented various large-scale land-use zones. The most common colour on all three combined maps is yellow. Prost employed it to depict 'Terres [et] terrains non bâtis.' The large expanse of yellow undeveloped land contrasts with the area's green space, red industrial zones, and orange housing. With it, he identified potential new construction sites and further confirmed to the reader knowledge of the region's terrain. The new 35-kilometre Parisian perimeter would provide ample room to accommodate more newcomers. Territory adjacent to the city of Paris itself contains very little of this vacant yellow land. However, readers can locate some of it peppered throughout the suburbs in close proximity, no doubt from the region's increased *lotissements* as the government continued to divide up large parcels of land in the early twentieth century.[14] Unsurprisingly, buildings occupy most of the property closest to Paris, but more open land becomes available and monopolizes the rest of the region as the reader moves further out towards the periphery. Between 1920 and 1935, area developers constructed on almost 2,000 hectares of open land each year, even though many of these sites were often no more than *lotissements défectueux*. Throughout the 1920s alone over

14 Soulignac describes the effects of the *lotissement* process as 'assez négatifs,' due to 'les conditions défectueuses dans lesquelles ils se développèrent' (195). Yet the process affected a large portion of the region's area, and allowed for much building.

185,000 underdeveloped plots appeared. In them, 504 subdivisions could be found in the Seine department (Paris and suburbs), 1,625 in the Seine-et-Oise department, and 250 in Seine-et-Marne, housing almost 700,000 people (Bellanger 297). The government paid little attention to these residents until parliament passed the loi du 19 juillet 1924, which gave official status to existing *lotissements* and allowed the government to regulate them. However, Minister of the Interior Camille Chautemps read the law as only 'faire application de ces principes aux lotissements futures ou actuellement en cours d'exécution.' As a result of this perspective, the loi du 19 juillet would target only future development and the government had no mandate to reform any current *lotissements défectueux* retroactively (Bisson 366). This interpretation did little to help the large numbers of residents who lived in existing shanty towns. For any significant betterment of those substandard plots of land in the region, planners would need to know them. Once acknowledged, agencies and organizations, such as the Musée social, could start to improve them.

Orange, the second most common colour found on the *Carte générale*, documents the region's 'Habitations.' Residential developments represent an important zone for leaders and citizens. Substandard suburban housing was one of the principal reasons for the 1924 legislation and Prost acknowledged this situation by documenting where the region's suburbanites currently lived. For the most part, 'Habitations' equate to already established jurisdictions. However, since Prost employed the army's *état-major* map as a base for his *Carte générale* and coated it with his own colour choices, his darker tints – in this case, orange and red – cover over most of the existing names. Via this presentation style, Prost hid much of the army's nomenclature. This allowed him to choose which names to keep and which ones did not matter enough to do so.

Prost labelled a few villages and towns, whose selection emphasized his view of their importance and worth to both planners and the nation. Many of these *communes* had historical, political, cultural, or economic roles in the region. Saint-Denis is one such example, along with Versailles and Saint-Germain-en-Laye, both of which Jaussely himself made visible to his readers. To the east of Paris, along RN 34, Prost marked Chelles. Under normal circumstances, this choice of a small outlying village would appear counterintuitive on a map promoting the future of the Paris region, but President Poincaré experienced trouble there in 1928 and, in many respects, Chelles helped restart the greater Parisian regional urban debate. Elsewhere on the map, Prost identified Argenteuil, the once-sleepy artist village on the banks of the Seine that joined with other suburbs at the end of the nineteenth century to form a border of development to the city's north-west and became a miniature manufacturing hub. The town's proximity to some of the earliest rail lines built in the region, from the Gare Saint-Lazare on to other points north-west, brought new residents and added to its role in the urban conversation. Obviously well-known to this map's readers, this *commune* appeared on the Service géographique's *Carte de France*, Jaussely's *Plan d'extension*, and now on Prost's *Carte générale*. Prost also incorporated Maisons-Lafittes, Asnières, and Nanterre into his cartographic discussion. Workers at factories in these towns, as in many others, lived near their employers and these *communes* informally 'close' an industrial loop to the north-west of Paris. Immediately adjacent to Paris to the north-west, Prost documented one of the largest clusters of suburban *habitations*, representing a historically prominent industrial area.

To the east, just north of the Bois de Vincennes, red can be found around Montreuil, Bagnolet, and Fontenay-sous-Bois, whose populations were not dissimilar to those Haussmann evicted in the 1860s. Aside from these few large clusters of residential settlements, Prost also includes numerous smaller orange towns and enclaves scattered throughout the region. Parisian expansion had completely transformed once-rural villages. Furthermore, they now participated in the Greater Paris urban conversation by default.

Albert Guérard tried to explain the disjointed growth of towns and villages in the region around Paris and suggested a solution. For him, 'Paris' was not just the city itself, but

> l'ensemble de ceux qui vivent de sa vie: c'est-à-dire qu'il comprend toute sa banlieue industrielle, maraîchère et résidentielle, [...] Il projetait autrefois des tentacules le long des routes: aujourd'hui, il les lance plus loin encore sous la forme de voie ferrées, avec des nodules ou des ganglions qui sont les gares. Une bonne carte de la propriété bâtie, et surtout de la densité démographique, révèle cette configuration extrêmement irrégulière, et pourtant logique, de la masse urbaine. [...] La création des satellites demandera, croyons-nous, l'intervention

> des pouvoirs publics. [...] Pour déclencher le mouvement [vers des villes satellites], rien ne serait plus efficace que la collaboration sur un point soigneusement choisi par la Ville et d'une grande compagnie immobilière. Organiser un centre modèle, avec transports rapides et tous les organes nécessaires à la vie civique, serait une entreprise grandiose et d'un rapport certain. Nous voilà bien loin des lotissements, bons ou mauvais, qui, depuis un quart de siècle, créent autour de Paris une vaste ceinture de laideur et de lésine. (qtd in Bellanger 203–04)

The idea to create satellite communities linked to the centre of Paris by transportation and built around shops and services, was regarded as a potential answer to the current large band of unregulated slums. Guérard felt that the government should plan and build them. After the Second World War, Paul Delouvrier and the Schéma directeur d'aménagement et d'urbanisme de la région de Paris proposed various *villes nouvelles* with this same composition and goal. The use of colour to expose the haphazard placement of many of these communes allowed Prost to remind readers of the disappointing results that when individual suburbs tried to plan and regulate themselves independently. Currently, Greater Paris's 'Habitations' extended to Arpajon and Corbeil on the southern edge of Plan III, to Creil and Verberie to Paris's north, as far west as Manté and Montainville, and to Villeneuve and Treilbardon on the eastern edge of the region. These villages, often 'foreign' to many Parisians, now appeared as members – albeit distant ones – of the new 35-kilometre region. Their inclusion affirmed that the region was composed of a variety of peoples with a multitude of needs and expectations.

The red of the 'Usines, terrains industriels, voies ferrées' is perhaps the most visible land category to identify on Prost's *Carte générale*. A vital component to any large, early twentieth-century metropolitan area, French industry contributed greatly to the region's economy. Greater Paris, home to a significant percentage of the nation's manufacturing capacity, had attracted workers throughout the nineteenth century. In the newly designated *région parisienne*, there were 11 *pôles d'emplois*, that generated the most jobs and experienced significant population growth. In Ivry, for example, to the city's south-east, 'les migrations sont intenses [en 1931], puisqu'on compte 9 461 départs, et 10 571 arrivées, chaque jour' (Soulingac 85). On the map, readers also see other *pôles* such as Montrouge, Suresnes, and Le Bourget. On the *Carte générale*, Prost prominently displayed these large industrial complexes, many of which abut orange *habitations*. Via his categorization of this sector of the economy, Prost documented the large proletariat populations that normally encircled manufacturing compounds around the city. But even if this political class now had a presence in the Chambre des députés, the categorization of all territory in the region was a critical piece of the new legislation. Pierre Laval's conservative government may not have wished to give official map space to the radical left, but Prost recorded these zones accurately.[15] As he did with towns and villages, Prost's sparse labelling on his map helps accentuate his views. He did not attempt to name all the industrial zones, a choice that helps curtail their potentially troublesome nature. With no name or label, most factory environs become another red blot on the map and blend into the greater suburban surroundings, indistinguishable from all the others. Just to the north of Paris, towards Saint-Denis, for example, Prost covered the names of the large red industrial *pôles* of Saint-Ouen and La Courneuve.

Prost removed their identity and power. He also reduced the potential for these communes to project disruption. Like Haussmann in the previous century, Prost divided up these troublesome areas. The Second-Empire *préfet*'s wide boulevards broke up working-class neighbourhoods in the hope of rendering future barricades all but impossible. Prost's 1934 limited-access roads would do the same in the twentieth century: divide up large segments of the working poor and disrupt them enough to limit their ability to organize and revolt. The same roads that divide these communities would allow car owners (the bourgeoisie) to pass through them easily.

To the north-east, beyond Paris's nineteenth *arrondissement*, readers clearly see the *abattoirs* and *marché des bestiaux*, another large cluster of industrialized terrain along with accompanying rail lines. Rail, an important mode of transportation, complements roads, work,

15 Pierre Laval was Président du Conseil des Ministres when the 1932 legislation was being debated. Although originally elected to the French parliament in 1902 as a member of the Société Française de l'Internationale ouvrière (SFIO), the French socialist government, Laval later served in most of the right-wing governments in the early 1930s and under Pétain in the Vichy government during the Second World War.

and manufacturing. To the south-east of Paris, beyond the Halles aux vins and the Bercy warehouse complex, which had a significant cartographic presence in 1889 and 1906, a third significant collection of industrial sites near Ivry has now expanded. Prost again covered up its name, so its proximity to Paris's border provides access to goods but only attentive or diligent readers will know its name.[16] Of course, all of these settlements contributed jobs and taxes to the French economy at a time of hardship, so they should not be ignored. Cartographically, they existed and their colour identified them. Prost's presentation, however, (few names, bright colours), minimizes these potentially problematic populations. He made known that industry, traditionally 'silenced' as exemplified in 1889, now abutted the city, but in a fairly benign way. With few names present, however, he removed much of their power. Most drivers leaving Paris for the provinces would have to traverse some of these zones. At some point, however, the construction of new limited-access roads allowed Prost to show them how to do so with minimal inconvenience. With this information at their disposal, bourgeois automobilists could either bypass these communities altogether or traverse them with minimal contact and avoid any real or imagined danger.

Green Parks

Roads, housing, and factories all played an important role in the region's economy. Yet the significant amount of green space on Prost's Plan II [East-West region] demonstrates his desire to document parks and woods and embellish the region's overall appearance. Throughout the late nineteenth and early twentieth century, scientists and urban planners, backed by the Musée social, tried to 'open up' cities to bring in sunshine and fresh air. Haussmann's Parisian boulevards were meant to transform the dark, damp slums, home to many cholera outbreaks throughout the 1840s, into safe manageable zones that residents could enjoy. Throughout the Second Empire and Third Republic, governments built, expanded, and maintained parks, all for the betterment of urbanites. Improved

16 Albert Demangeon described 'le gros centre industriel' of Ivry, containing '[les] industries métallurgiques, industries chimiques, industries céramiques, [et] industries alimentaires' (54).

transportation, in the form of omnibus, subway, rail lines, and private automobiles further increased citizens' access to both green space inside the city and woods outside Paris. The Musée social's urban and rural hygiene section initiated, along with the Alliance d'hygiène sociale, the Société française des habitations à bon marché, and the Société d'art populaire, a movement to promote the cause of *espaces verts* in conformity with the standards of public health (Horne 250). As in the maps produced by Alphand, the Service géographique, and Jaussely, the Bois de Boulogne and the Bois de Vincennes continues to dominate the city's park system in Prost's plan. Their size and proximity to Paris offered fresh air and space for relaxation to generations of city dwellers. Outside the city, other large wooded areas now provided many of the same benefits to the increasingly developed suburbs. Expansion during the 1920s, however, meant that these woods, forest, and fields were disappearing fast. One of the Comité supérieur's fears was the complete loss of the region's last great forests. The grounds near Versailles and Saint-Cloud, Marly, or Sceaux, along with the forests of Meudon, Verrières, Malmaison, Cormeilles, and Saint-Germain-en-Laye, all represented part of France's national heritage and needed to be saved from urbanization (Delorme 24). Their inclusion here suggests that Prost agreed with the committee and enlarged available green space on the map/in the region.

On the *Carte générale*, large woods form a forested arc from east of Versailles to north of Saint-Germain-en-Laye and create a *de facto* green belt west of the city, an effective buffer against further industrial or residential development in this previously agricultural zone.[17] In an effort to avoid potential crowding and to maintain a semi-rural feel in the outer zone of the region, Prost 'projetait d'adjoindre à ces espaces verts [existants], ainsi que l'on dit en jargon d'urbaniste, les surfaces agricoles d'étendant, au Sud et à l'Est, aux portes mêmes de Paris, en les transformant en parcs, jardins et terrains de sport à proximité

17 Demangeon describes the band of territory from Neuilly to Clichy as 'une belle ville, luxueuse, aux larges avenues bordées de maisons cossues, qui forme comme le prolongement des riches quartiers parisiens.' However, he also points out that 'en face du Bois de Boulogne, sur l'autre rive de la Seine,' in once-bucolic agricultural *communes*, such as Suresnes, 'commence une zone industrielle qui a gagné Gennevilliers en passant par Puteaux, Courbevoie et Asnières' (58).

de quartiers déshérités' (Bastié, *Urbanisme* 161). Just as Haussmann's transformation of the Parc des Buttes-Chaumont in north-east Paris from an infamous dumping ground into a highly prized park, Prost suggested an enhancement for the eastern suburbs, especially those disadvantaged communities without promenades or recreational areas. The concentration of green would raise land values and improve the lives of those who lived nearby. As if to confirm a higher population density than found in the bourgeois areas on the other side of the metropolitan zone, the 'parcs, jardins, terrains' here, if built, would be smaller than their western counterparts but nonetheless serve an important function in the area. North of the Bois de Vincennes, which 'a perdu sa tranquillité; les arbres ne l'occupent plus en entière de larges clairières' (Demangeon 51). Near Montreuil and Bagnolet, for example, Prost recommended the creation of a small green arc, no doubt a welcome addition to the collection of working-class enclaves near Noisy-le-Sec, where one encountered 'le bruit d'une grande gare [et] l'intense trafic de la ligne de Mulhouse, de la ligne de Strasbourg et de la Grande Ceinture' (Demangeon 51). With available vacant land at a premium, developers prized any open space near Paris. The post-1919 housing debate between developers and the Musée social took on a very contentious tone as the former pushed for construction on a larger part of the open space of the fortifications. Yet on the *Carte générale*, lands adjacent to the former walled fortifications are green and presented as 'emplacements destinés soit à des promenades publiques ou terrains de jeux, soit à des services publics, soit à des cités jardins' (Bastié, *Nouvelle histoire* 236). In that spot, Prost suggested the placement of both green parks, heavily championed by the Musée social, and a *boulevard périphérique*, desired by the government and men in power. The 35-kilometre ring around Paris, a 'wasteland notorious for ramshackle shantytowns,' was home to an estimated 42,000 people by 1926 (Horne 251). By 1939, 38,750 housing units had been constructed there, mostly in the form of 'grands immeubles collectifs' (Bastié, *Nouvelle histoire* 238). By clearing these residents, Prost provided what the Musée social had long desired: a ring of parks and playing fields around the city instead of unregulated shantytowns.

Although Prost documented and coded development with bright and noticeable colours, he left a significant portion of his document in black and white. The largest amount of grey territory, which would not be modified or studied by him, lies outside the newly designated *le grand Paris* (the area within the 35-kilometre radius from Notre-Dame), and would not benefit from Prost's ideas. Until any community beyond this geographic boundary reached the 10,000 resident threshold set by the 1919 loi Cornudet, it must either turn to a department or to the national government for any large-scale infrastructural improvement. The *communes* of Melun to the south-east and Meaux to the east are perhaps the best-known (and visible) settlements outside the new official region. The remaining grey territory on all three plans is peppered with smaller, lesser-known towns and villages, representing potential barriers if the government felt that future enlargement of the Paris region was needed.

Perhaps the most conspicuous non-coloured space on the map is Paris itself. On this map designed to improve 'l'aménagement de la région parisienne,' the city is completely absent on both Plan I (région nord) and Plan III (région sud). If Prost were to publish those sheets independently, a potential reader may not recognize them as part of the Paris region. On Plan II (région est et ouest), however, Paris is visible, labelled in all capital letters, affirming Prost's desire to acknowledge the city's magnitude and its central position in the region. Yet, shockingly, he has removed its colour. In doing so, Prost presented the French capital as a lifeless space surrounded by a vibrant, colourful, growing suburban ring. With this portrayal, Prost encouraged readers to focus their attention on growth beyond Thiers's former wall, but also confirmed that the embellished capital city was 'complete.' Government officials wished to keep Paris out of the urban debate for the time being, as their predecessors' had both the time and cost of Haussmann's works.[18] Construction outside the city, however, also meant much less disruption for Comité supérieur members.

In terms of *intra-muros* transportation, there is little to note. With the exception of two red arrows that puncture its outer skin, the bright-blue Seine, and the Canal Saint-Martin flowing through the city, Prost's Paris has no contact with the outside, suggesting yet again

18 In 1869, Haussmann floated a loan of 250 million francs, which he had hoped to repay in ten years out of the normal city budget. But his fall and the war changed that. After the war, the city owed 313 million francs. By December 1879, the city council passed a resolution to the effect that it would never again finance public works by borrowed money, but would pay for them directly out of budget surpluses (Sutcliffe 55–58).

that only the suburbs needed the government's attention. Although home to a sizeable percentage of the region's population, its grey representation negates the capital's vigour. Prost affirmed that all roads led to Paris, but any construction or infrastructure upgrade would stop at its border. Even as development in the suburbs flourished in the early 1930s, Paris's former walled fortifications, now gone, remained a mental barrier, a stubborn, fixed divide between city and suburb.

Bright colours draw the reader's eye to what was important: roads, industry, housing, and green space. Goals that seem to have eluded national and city officials since before the Revolution. Where Alphand's 1889 map highlighted new street construction and Jaussely's 1919 project showed how much green space the Paris region could offer, it fell to Prost to reconcile many of the shortcomings that plagued the growing region. He conformed to the 1932 legislation and provided a pathway to improving Greater Paris. His *Carte générale* presents a brightly coloured expansive urban agglomeration, one that fully embraced the new 35-kilometre legislative boundary. His ambitious plan exemplifies the changing notion of cities and how people use them.

Conclusion

On Bastille Day 1935, one year after the publication of the *Carte générale*, 30,000 conservative Croix-de-feu supporters demonstrated in the wealthy western districts of Paris against the rise of the left. On the city's poorer, eastern end, however, 'unruly crowds of Radicals, Socialists, and Communists paraded three hundred thousand strong – ten times greater – as proof that change was under way' (Martin, H 39). France's parliamentary elections in May the next year produced a clear majority for the left-wing Popular Front, and ended the conservative government of Pierre Laval, when it took 380 of 608 seats in the National Assembly. France's new 'working-class' majority had come to fruition in large part due to significant population growth in Paris's *ceinture rouge*, the same area highlighted on Prost's map. The Paris region's suburban inhabitants helped elect this coalition of the SFIO, le Parti radical-socialiste, and le Parti Communiste led by Léon Blum, and placed France's first Socialist Prime Minister in Matignon Palace. Although others, such as Guérard, documented the demographic shifts outside the capital, Prost's plan, published just two years prior to this upheaval, predicted the rise of this political demographic by revealing the true size and complexity of the suburbs. 'Rediscovered' by Poincaré in 1928 and highlighted by Prost in 1934, Paris's suburban working-class's introduction to the city's political elite was completed by the Popular Front's victory, albeit short-lived, by affirming their political power through this remarkable victory at the ballot box.

When Prost presented his *Carte générale* to the Comité supérieur two years earlier, on 14 May 1934, they viewed the newly created enlarged administrative division cartographically for the first time. Prost's map provided them with both a comprehensive view and a new understanding of what was happening past the former walled fortifications. Throughout much of the city's history, mapping the French capital has meant focusing the reader's attention on Paris itself, a practice that prevails in large part today. This preference also mirrored where the government had traditionally targeted its resources. However, as a language of power, historical maps also allow contemporary researchers to uncover evolving discourse shifts. On Prost's *Carte générale*, for example, Paris's urban core loses its cartographic power as he shows the rising prominence of the region's periphery.

Subsequent legal delays and consultations, initiated primarily by defiant suburban communities, suspended implementation of the *Plan Prost* for five additional years and its eventual enactment, by government decree, did not take place until 22 June 1939 (Bastié, *Urbanisme* 117). Nazi Germany's invasion of Poland a few months later definitively ended any chance of Prost's project becoming a reality. In one last gasp of hope, on 28 August 1941 the Vichy government used Prost's plan to create PARP (Projet d'aménagement de la région parisienne). Defeat and the subsequent German occupation prevented France yet again from initiating sweeping urban reform in Paris or elsewhere. Yet, in many ways, the *Plan Prost* remained relevant following the Allied victory in 1945. Just as he foresaw the rise of the suburb, which created a fertile ground for the Popular Front's pre-war rise, Prost's *Carte générale* anticipated a post-war agglomeration where limited-access *autoroutes* and accompanying highway interchanges cultivated the individually owned automobile's domination of the suburban cityscape and kept the embellished city centre free for business and leisure.

The 1920s and 1930s brought considerable change to residents of the Paris region. Significant social shifts, manifested through electoral victories and new legislation, enhanced the way of life for a previously neglected class of citizens. An early twentieth-century address beyond Paris's former walled fortifications most likely meant either a comfortable individual house in a village or bourgeois community or a small space in one of many crowded shantytowns fabricated poorly with low-quality materials and little or no regulation. As undesirable as they may have seemed, those working-class suburban communities continued to grow, fed by newly arrived residents in search of cheap housing and proximity to factory jobs. At the time of president Poincaré's experience in Chelles in 1928, few top-level politicians had a good understanding of the quality of life there, but most agreed that the *banlieue interminable* could not and should not go on forever. The loi du 14 mars 1932 laid the foundations for comprehensive, yet ultimately unrealized, suburban reform.

Planning for Greater Paris meant that officials needed to understand and visualize the region. Used to 'looking' at land outside of the city almost exclusively on maps, such as the *carte d'état-major*, the *Carte de France*, or Jaussely's *Plan d'extension*, the 'view' of the suburbs from administrative offices in Paris itself was not accurate or up to date. Akin to de Certeau's view from atop the World Trade Centre, Henri Prost's final product, more specifically his 1934 *Carte générale* represents the view from above, where the busy city below appears calm and manageable. Of Paris's officially sanctioned Third Republic maps, Prost's comes closest to acknowledging the *là*, providing an overall view of the entire region and the *ici*, as it acknowledged, surveyed, and documented Paris's chaotic suburbs. Perhaps most importantly, Prost's map demonstrated to the Comité supérieur that, beyond recognizing the region's outer boundary and drawing new freeways, the French capital had a new identity, to which all the region's residents now belonged.

In more practical terms, Prost's map affirms Guérard's analysis of the Seine department's demographic growth and demonstrates how the region's expansion would eventually transfer some power and influence away from central Paris. Through graphic components (colour and symbols), Prost's map discourse displays this historical shift more effectively than Guérard's population statistics alone.

The Paris region today continues to experience the same dichotomy developed throughout much of the city's history. High housing costs continue to push lower-income families further away from the city's embellished centre towards the outer suburbs, where an insufficient transportation network necessitates long commutes and an isolated existence. Perhaps most telling of the ineffectiveness of this divide, one could easily use the same vocabulary employed by Poincaré in 1928 to describe the massive *habitations à loyer modéré* (HLM) constructed in the city's post-war suburban industrial communities. Prost's *Carte générale* sounded a warning for pre-war Paris. If they refused to address and enhance the *ceinture rouge*, generations of suburban residents, and the capital itself, would lose.

Conclusion

The German invasion of Poland in September 1939 precluded any urban reform for Paris, or the rest of France. The Nazi occupation of the capital and northern half of the country the next year meant that the Vichy government, not republican politicians, would now decide Paris's future and it did not take long for the new regime to assert its control. On 14 October 1940, the government suspended the Conseils généraux. In doing so, it removed elected officials from the department-level governing process. The loi du 16 novembre that same year gave Vichy officials the power to name mayors and municipal council members in *communes* with more than 10,000 residents. On 2 February 1941, the government suspended the 1932 Comité supérieur d'aménagement et d'embellissement et d'extension de Paris and created the Comité d'aménagement de la région parisienne (CARP) under the direction of the Délégation générale à l'Equipement National (DGEN). As part of this new mandate for Parisian administration, the Seine *préfet*, named by Vichy, had the power to 'proposer toutes les modifications qu'il jugera nécessaires d'apporter aux plans des communes de banlieue, pour les subordonner à celui de Paris' (Cohen 236–37). With this decree, the Vichy regime returned to a Haussmann-style executive, whereby one man had the means and power to decide what was best for Paris. Over the next four years, Vichy hoped to 'reimagine the decaying monster of Paris as a capital worthy of French greatness' (Wakeman 297). In order to achieve this sought-after prominence, the *préfet* would have to fix many perennial problems. One primary concern was new construction to house internal migrants who had arrived during the Third Republic. However, under the German occupation, a skyscraper-centric New York or Chicago-style skyline, the antithesis of French modernism and home to the declared enemies of the Vichy authorities, would not find a favourable reception in the French capital. The June 1944 Allied landings on the Normandy beaches terminated any chance of Vichy officials or their German counterparts influencing Paris urbanism for the long term.

With the end of the German occupation and the need to rebuild not only Greater Paris but much of the northern and eastern sectors of France, it fell to the national government to dictate the path forward. Reconstruction occupied the bulk of the French urban discussion in the first few years after the war. However, by the spring of 1950, France was ready to plan its future. The Institut de l'Urbanisme de Paris's journal *La Vie urbanine* published an article entitled 'La Région parisienne' in which the author, Pierre Gibel, Chef du Service de l'aménagement de la région parisienne, discussed the development of the Paris region from the 1793 *Plan de la Commission des Artistes*, with which the revolutionary authorities planned to renovate the city with *grand axes*. How could urbanists, planners, and bureaucrats fix the region's problems in a manner that benefited the largest number of residents? For his 1950s article, Gibel focused his attention primarily on the suburbs, where he felt that 'la tâche d'aménager la région parisienne s'est trouvée accrue par la nécessité de procéder en premier lieu à la reconstruction des soixante-dix communes sinistrées de la région' (135). Post-war rebuilding provided a unique opportunity to demolish unhealthy and unsafe housing, and to clear the way for new construction in many areas that would have been politically difficult before. Improving the insalubrious suburbs could provide maximum benefit for the neediest segment of the population but also a government looking to gain favour with working-class voters. Furthermore, an ordered suburban ring to accompany the Haussmann-style modified city centre could help eventually erase the centre-periphery divide.

Documenting the Region

After the war, the capital region's needs overshadowed much of the national urban discussion. This prioritization continued the convention whereby the modern urban centre dominated the tradition-bound periphery (Sawyer 20). An additional underlying post-war goal therefore would be to help cultivate the margins. The betterment of these spaces, regularly overshadowed by Paris, could demonstrate progress in the government's desire to improve the lives of all citizens. Long the recipient of the government's attention, the city had monopolized the nation's resources for centuries. As the political, commercial, and industrial capital, Paris occupies a central position in the nation's consciousness. Its growth and development, often at the expense of the provinces, siphoned away people and money needed by the rest of the country. This dichotomy appeared at the regional level too, where the expanding edge that housed most of Paris's newly arrived working-class domestic migrants contrasted greatly with the embellished bourgeois city centre. Map space, always at a premium, serves as visual proof of the rise of these suburban communities around the turn of the century, when the previously silenced periphery began to eclipse the city itself. In 1889, Alphand's bias towards the bourgeois centre all but ignored the territory beyond Thiers's wall. But 45 years later, Prost instead privileged land outside Paris as he documented his view of a future *région parisienne*.

The physical composition of post-war Greater Paris had changed remarkably since the Revolution's 1793 *Plan de la Commission des Artistes*, but the city maintained its role in French politics and the national economy. Paris accumulated additional power and influence throughout most of the nineteenth century. Napoleon concentrated the national bureaucracy and administration there as he professionalized French nation building. Under the July Monarchy, an 1841 law mandating that rail lines radiate from Paris, combined with the Second Empire city's growing role in the world's financial markets, continued that momentum. At the same time, Parisian urban space retained its negative connotations as its citizens overturned governments in 1789, 1830, 1848, and 1871. Yet the city maintained its economic draw. Industrial advances throughout the nineteenth and early twentieth century lured rural citizens to the city in search of jobs in factories and a better life. Unfortunately, Paris could not accommodate all of France's domestic migrants. These newcomers contributed to the nation's economy and the region's population growth, but also to the insalubrious state of both society's and the city's edge.

The cartographers examined in this book recorded how the physical and mental divides between the geographic centre in Paris and the growing economic and physical periphery faded throughout the Third Republic. Between 1889 and 1934, maps focusing on Paris covered more territory but the percentage of map space, perhaps the definitive determinant of an area's power or influence, devoted to intramural Paris contracted. Although Vichy viewed urban space with contempt, the regime failed to put an end to the nineteenth century's Parisian centralization. Gibel's article in *La Vie urbaine* further affirmed this trend and he explored the region's progress in the previous century and described the current state of the post-war agglomeration. By 1950, the Paris region, home to one-seventh of the French population but only one-fortieth of its territory, continued to play an outsized role in nation's political and economic arenas.

To help illustrate the current state of post-war Paris, Gibel included several basic cartographic images, all of which draw readers towards the outer periphery.[1] The *Carte administrative*, taken directly from the Ministère de la reconstruction et de l'urbanisme, shows a simple outline of the official post-1943 *région parisienne*, which included the Seine, Seine-et-Oise, Seine-et-Marne, and the southern cantons of the Oise departments. On it, a dark 'zone centrale de forte densité' covers both Paris and the inner suburbs. A lighter 'zone de faible densité' includes outlying communities (116). Gibel's cartographic choices encourage his readers to both question and reimagine the old centre. Under Vichy, the political make-up of the Paris region had expanded well beyond the 1932 legislative 35-kilometre limit and the Reconstruction Ministry's map demonstrates the extent to which suburban growth now superseded that of the urban core. In addition to showing readers current and accurate jurisdictional boundaries and their relation to the city of Paris, Gibel also included a *carte physique* to outline the topography of the Paris Basin and document the physical base, upon which city and suburbs could grow.

Post-war France still hoped to invest in Paris's suburbs. The proclamation of the Fifth Republic in 1958, the return of Charles de Gaulle, his centralization drive, and his distinctive sense of France's grandeur among nations, saw the Paris region's spatial production increasingly corralled and organized into a concerted programme of state-led

1 The point of these maps seems to be more to briefly highlight where the state wished to target development in the Paris region. Other maps include 'le schéma routier,' 'les établissements d'enseignement du second dégrée,' 'les principaux aéroports,' and 'les zones de cultures maraîchère et fruitière qui doivent être préservées' (Gibel 141–47).

development (*aménagement du territoire*). Although national spatial planning was overseen by the DATAR (Délégation à l'aménagement du territoire et à l'action régionale), created in 1963, development of the Paris region was the responsibility of Paul Delouvrier, a high-ranking administrator who had served as the Gaullist government's representative in Algeria between 1958 and 1960 (Welch, 'Passagers' 177). De Gaulle charged Delouvrier to instil order over the chaotic periphery that grew outside the city limits.

Maps as Control

For much of the previous century, the evolving relationship between cities and the countryside bordered on the confrontational. Paris represented a 'center of power, domination, and opportunity. It was both envied and hated' (Harvey 316). Many politicians from across France targeted the capital. Troubled by nineteenth-century uprisings, these men believed that urban centres and their inhabitants represented what was wrong with France. The political and social agenda of *les ouvriers* conflicted with traditional bourgeois ideals and goals. So, map-makers preferred to silence poorer neighbourhoods and their residents, robbing them of a visual cartographic presence as well as a political one. By doing so, they ignored narrow alleyways and shantytowns and left blank the zones they occupied on maps. This blatant lack of information created unfamiliar spaces and deterred bourgeois residents from venturing into extramural communities or even intramural urban neighbourhoods, like Belleville, which many of them considered 'foreign territory' (Harvey 317). In the 1920s, the ever-growing 'other,' those residents who lived beyond Thiers's wall, started winning elections and amassing political capital. Universal male suffrage provided the working class with a stronger political voice and this electoral power showed up on maps. Jaussely's 1919 *Plan d'extension* recognized that the suburbs needed to be addressed and the 1924 legislative and 1925 municipal elections that created the term *ceinture rouge* proved him right. By the 1930s, Prost too acknowledged the enhanced role and draw of the suburbs.

With social and political reforms in motion, the Third Republic eventually calmed Paris and invited its residents into the larger political discussion. This stability enhanced the city's role as a destination. The arrival of both domestic and international tourists to see Haussmann's wide boulevards and the Expositions universelles in 1878, 1889, and 1900, showed how Paris's urban space could benefit the nation. Furthermore, other large-scale events, such as the funerals of Adolphe Thiers and Victor Hugo, proved to officials and residents alike how effectively the Republic could control Paris's crowds, and that the city was no longer something to fear. Maps affirmed cartographic control.

In addition to political shifts, the individual urban experience also changed. Technology perhaps did most to alter how residents related to each other and to an expanding urban footprint. The introduction of intercity and intraregional rail, and later the automobile, meant that residents and visitors alike could interact more easily with people from different social classes and geographic locations, but also with the city, region, and nation very differently from generations prior. Commuting longer distances became easier as the time needed to reach central Paris from outlying *communes* shortened. These changes slowly altered their mental definitions of urban space. What the government or a private citizen considered 'Paris' now encompassed a much larger territory than it had before the Third Republic. The increased migration to the suburbs from Paris led to considerable building and expansion outside the traditional core. Yet the dense centre, although now within easy reach from the farthest suburb, still contrasted greatly with the spread-out nature of disparate and often-neglected suburban communities throughout the region. To unite the two effectively, post-war Paris had to plan on a larger scale. In 1967, the Institut d'aménagement et d'urbanisme de l'Île-de-France published a photo of Delouvrier sitting with the *préfets* of Paris, Seine-et-Marne, and Seine-et-Oise in front of an aerial photo of the Paris region (Welch, 'Carte et territoire' 193). Although Third Republic cartography provided the requisite foundations for future twentieth-century urbanists and planners, Delouvrier took advantage of satellite technology to furnish his view from above. Yet, the premise of Parisian bureaucrats trying to better understand the changing territory below in order to impose order on residents there, remained very much the same.

The Democratization of Cartography

As the government used maps to document the changing Paris region, more people saw and understood them. Prior to the Third Republic, map reading remained limited to a restricted population, primarily people with the education and means to understand them. Jules Ferry's 1880 reforms helped improve map access significantly. They provided mandatory, free, and secular instruction to millions and the tools to understand and participate in the new democratic form of government. As part of this educational drive, geographers, such as Vidal de la Blache and Jules Simon, pushed to expand the place of their subject in the national curriculum. To assist in this mission, these men helped get cartography into classrooms via wall maps, textbooks, and teacher training. With this subject now being taught to future businessmen, bureaucrats, and soldiers, prospective French leaders could see what the nation looked like and turn to geography to validate France's existence. This interaction ended a cartographic void and allowed the government to use mapped presentations of France to craft a republican narrative whereby its people, united by a common language, history, and now geography, could support it more enthusiastically as the nation looked to regain its rightful position within a post-1871 Europe.

Students' daily exposure to cartography via wall maps and textbooks played a huge role in normalizing the medium. Fouillée's *Le Tour de la France par deux enfants* championed republican values and introduced various corners of the nation to generations of school children. As the protagonists circled the country, they both epitomized and highlighted French values. The two boys experienced the advantages of speaking one common language, witnessed how hard work could improve the lives of their fellow compatriots, observed the dangers of too much drinking, and promoted the benefits of a well-rounded education. André and Julien spend the year learning about French history, France's economic strengths and scientific innovations, and Fouillée situates every lesson within a region or department of the country. Starting with the 1884 edition, she also accompanied the boys' adventures with regional maps. This cartographic inclusion could only happen due to the increased role of geography in classrooms around the country. Furthermore, Fouillée's decision familiarized map reading for millions of school children as it gave them the opportunity to view, interact, and trace her cartography on a daily basis.

Maps also helped update the nationwide geographic vernacular. Citizens used to referring to their old provincial region, such as 'Le Berry' or 'Le Dauphiné,' could still find them in Fouillée's narrative, even though a 1789 decree and 1790 implementation terminated the official use of such royal designations. Even as late as the Third Republic, titles for the republican-established departments, relatively uniform in size and named after rivers, mountains, or other geographic components, had failed to replace the former provinces in daily speech. Fouillée's inclusion of provincial names tells readers today that teachers and parents still used them more than departmental ones, but her maps and narrative contained both. The map of Brittany, for example, is titled 'La Bretagne' in the table of contents and in the legend. However, when looking at the map itself, a reader can identify the Finistère, Côtes-du-Nord, Ille-et-Vilaine, Morbihan, Loire-Inférieure, Maine-et-Loire, Sarthe, and Indre-et-Loire departments (231). Fouillée helps new map readers view and understand the actual administrative division of the nation. If a student's village was found in one of these eight departments, they could associate the two designations and see how it belonged simultaneously to both Bretagne and Ille-et-Vilaine.

As children and, subsequently, adults learned about and saw more maps throughout the nineteenth century, they also became more commonplace. By the First World War, cartography appeared more frequently in newspapers, travel guides, and public documents. Legislatively, politicians called on bureaucrats, and later urbanists, to publish maps and plans of the French capital city to accompany new urban regulations. Few would argue that the 1793 *Plan de la Commission des Artistes* had reached a very limited audience due to educational and accessibility shortcomings at the end of the Ancien Régime, but by the time Henri Prost published his three-page *Carte générale* in 1934, the *urbaniste-en-chef* could be confident in the knowledge that not only would a significant portion of the public at large be able to read and decipher his multisheet map, but also navigate its legend, and understand that they were looking at the Greater Paris region.

Professionalization of the Terrain

In addition to educational reforms, the diplomatic and scientific interaction between France and surrounding countries throughout the nineteenth century introduced new means of documenting the nation. The creation of the Napoleonic code unified France's legal system, highlighted the necessity for accurate information on French territory, and launched the quest to update the century-old, out-of-date Cassini map. Yet provincial academics played a role as well. Exposure to physical geography from Germany and central Europe inspired French geographers and cartographers, such as Vidal de la Blache from Eastern France, to focus on and improve their own topographic representations of the nation. In an effort to better know its terrain, the army commissioned its Dépôt de la Guerre, a professional corps devoted to understanding and documenting the national landscape, to survey and map its domestic territories. The army homogenized its presentation of French land, which made it easier for future citizens to read and understand its maps. It published the resulting versions of the *Carte d'état-major,* which allowed generations of French to see for the first time what many far-off departments and cities looked like. In addition, this standardization ensured that readers would recognize cities, roads, rail, parks, post offices, etc. on existing and future governmental maps, such as the Service géographique's 1906 *Carte de France*. In addition, this professionalization ensured that France's mapped product was accurate and up to date.

Map-makers championed their own point of view, but these biases also reveal how a society understood urban space. Through their cartographic discourse, maps in this book demonstrate the effects of social changes, technology, and population growth had to Paris. In addition, this cartographic diversity reflects not only jurisdictional but also technical progress. In 1889, to commemorate the centennial of the storming of the Bastille, Alphand's *Les Travaux de Paris* atlas celebrated the growth and improvement of the French capital in the previous one hundred years. His book champions the relatively new republic and its capital, which had recovered completely from war and Commune. His regime had invested heavily in all corners of the city via reds and yellows on Alphand's map 'Paris en 1889: les opérations de voiries exécutées entre 1871 et 1889' showed readers his yellow-and-red road construction. Thanks to the Third Republic, many more citizens would receive wide boulevards in their neighbourhoods. Yet, for its presentation of a growing city, the biased mapped portrayal of the heavily embellished city promptly ends at Thiers's wall. Bourgeois-centric bureaucrats felt no need to document any suburban territory.

With the advent of the Musée social and urban studies at the end of the nineteenth and beginning of the twentieth century in Europe, thoughts on how to view and experience French cities changed. The promotion of science and hygiene meant a greater investment in previously ignored *quartiers*. Following the First World War, government officials hoped to plan for the future and to instil order in its growing urban centres. The 1919 loi Cornudet required urban plans for any municipality with more than 10,000 residents. The legislation also allowed cities to acknowledge the new science of urbanism and apply it to where people lived. Instead of unlimited building and expansion as need and topography dictated, urbanists wished to control urban space in a manner that could most benefit residents. A city dweller needed parks and enhanced access to clean water. Furthermore, these new scientists incorporated recent technological advances in transportation, such as the personal automobile, aeroplanes, and subways into their projects. Paris opened up its urban planning to competition and Léon Jaussely, a member of the Musée social, won the capital's commission with his *Projet lauréat de la section générale du concours du plan d'extension de Paris*. His ideas, painted onto a Service géographique document, demonstrated what a modern French metropolis must contain in the near and distant future. He highlighted the movement of vehicles with limited-access roads. In addition, Jaussely acknowledged the importance of trade and commerce with the inclusion of a large port to the city's north-east as well as rail and industrial infrastructure. The future figures prominently in the 1919 *Plan d'extension* as well. He proposed various *aerogares* in all corners of the region plus two automobile racetracks. In addition to enhanced movement and pleasure, Jaussely suggested that the city place all cemeteries and industrial waste sites outside the city limits. Perhaps most important, Jaussely affirmed that 'Paris' was now more than just the city itself. On his *Plan d'extension de Paris*, the city itself appears full, with no room for growth. A majority of the cartographic space is devoted to the suburbs, where the most growth would take place.

The cost of repairing the extensive damage following the First World War proved too great and thwarted the Paris region's implementation of Jaussely's plan. Yet, following the cessation of hostilities, both the influx of new residents and construction on Paris's periphery continued. By the 1920s and 1930s this unencumbered growth meant more roadway gridlock and shantytowns outside the former walled fortifications into the outer suburbs. Legislation in 1928 created the Comité supérieur d'organisation de la région parisienne, one of the government's earliest attempts to integrate the suburbs into Paris's planning, but it was not until 1932 that the national government legislated a practical solution. The loi du 14 mars 1932 redefined Greater Paris. Instead of delimiting the region demographically, with a minimum of 10,000 residents, the legislation established the capital region's borders geographically. The new law required a comprehensive urban plan for the area that fell within 35 kilometres from the *parvis* of Notre-Dame Cathedral on the Île de la Cité. This expanded definition came thanks in part to the significant growth in the role played by suburban communities in the region.

Henri Prost, whom the Comité supérieur named as Urbaniste en chef of the Paris region, had two years to complete his vision for Paris's future. On 19 May 1934, he presented his *Carte générale de l'aménagement de la Région parisienne* to Albert Sarraut, Minister of the Interior. A disciple of the Musée social's principles, Prost continued many of Jaussely's ideas. For example, he championed ease of movement with several limited-access roads presented in a dark red. These *autoroutes* joined with *routes nationales* via *points de raccordements*, which mimicked the shape of traffic circles, allowed a seamless connection between the two levels of road, enhanced automobile movement, and allowed drivers to avoid problematic industrial areas. The stark colours of his roads stand out against the yellow of undeveloped land. In addition to potential growth zones, which could accommodate the region's demographic expansion, the inclusion of parks and existing *habitations* further clarify what existed inside the 35-kilometre regional boundary.

As the city drew less attention from the government planners, the size and shape of the periphery changed. The size of the new region, significantly larger than Jaussely's 1919 presentation, meant that Prost had to cover and document much more territory than his predecessors. Furthermore, the scale of his final map also demonstrates the breadth and size of the French capital. The new border had expanded to such an extent that Prost had to divide his map over three sheets, which added complexity to Prost's discourse and proved that the French map reader had become more sophisticated in his or her ability to understand intricate cartographic convention. Prost's goal was to sell his plan to the authorities. The map's immediate audience, members of the Comité supérieur, would be well versed in official French cartographic discourse, so a more specialized document made sense. Perhaps the composition of this audience also allowed Prost, in a complete reversal from other maps in this book, to remove colour – and therefore influence – from Paris, affirming the growth and influence of the suburbs. Maps illustrate this change better than numbers alone.

Although the Second World War stopped the Plan Prost and his *Carte générale* from moving forward, the map's contents testify to the changing nature of Paris along with the enhanced competency of the French map reader. The time and effort devoted to providing map access to the French throughout the Third Republic led to a competent functional interwar map reader. Furthermore, the mental views and understanding of Paris changed during this time. From a city limited to the space inside Thiers's walls to a three-sheet depiction of Greater Paris, the definition of a city and its components adjusted accordingly. Even today, the importance of transportation, housing, and green space remain essential for any urban setting. However, the eventual need to plan for the future provided leaders and residents alike with the means to project and order growth.

The ever-growing suburban ring perhaps represents the ultimate 'periphery' for French officials to document, understand, and conquer. Gibel's final section, 'L'avenir de la Région Parisienne,' perhaps best exemplifies the city administrator's views on the region's future. To solidify Paris's position as a world city, he wrote, 'il faut [...] oser prévoir grand. [...] La réalisation de l'aménagement de la région parisienne est une tâche de longue haleine' (152–54). Gibel showed how Paris grew and identified several 'opérations essentielles' that the region would need as it moved forward.[2] Enhanced cooperation between different departments and *communes* in relatively mundane matters such as potable water, trash removal, and a comprehensive electricity grid would help accommodate new residents and boost the capital's role in a post-war Europe and the world just as much as a new highway or airport. Of course, this debate continues and 60 years later, politicians are still trying to plan and define the region. On 1 January 2016, a new scheme, the Métropole du Grand Paris, came into existence. The idea of creating and administering a comprehensive and effective development project for Paris remains an important

2 Gibel mentions 'Circulation et Transports, Espaces libres, and Service publics' as a few of these essential operations (135–36).

goal. We can see the beginnings of many of the important aspects of Grand Paris, such as eased transportation, take hold as early as 1889 and develop more throughout the Third Republic. Even following the industrial decentralization initiatives from the 1950s and political ones under Mitterrand 30 years later, residents, business, and industry continued to arrive in the Île-de-France region. Yet administering the scope and population of today's Greater Paris will require a continued examination of all areas in and around the city.

Bibliography

Ageron, Charles-Robert, 'Les Colonies, élément de la puissance française?', in Pierre Milza and Raymond Poidevin (eds), *La Puissance française à la 'Belle Epoque'. Mythe ou réalité?* (Paris: Éditions Complexe, 1992), pp. 35–50.

Agulhon, Maurice, 'Le Centre et la périphérie', in Pierre Nora (ed.), *Les Lieux de mémoire, III. Les France* (Paris: Gallimard, 1992), pp. 2889–906.

——, 'Conscience nationale et conscience régionale en France de 1815 à nos jours', *Histoire vagabonde II* (1998), 144–74.

Akerman, James R, 'The Structuring of Political Territory in Early Printed Atlases', *Imago Mundi*, 47 (1995), pp. 138–54.

Allen, James Smith, *In the Public Eye: A History of Reading in Modern France, 1800–1940* (Princeton, NJ: Princeton UP, 1991).

Alphand, Adolphe, *Les Promenades de Paris* (Paris: Rothschild, 1868).

——, *Les Travaux de Paris, 1789–1889* (Paris: Ville de Paris, 1889).

Amalvi, Christian, 'L'Histoire à l'école et au lycée: les manuels d'Hachette (1830–1914)', in Roger Chartier (ed.), *Le Temps des éditeurs du Romantisme à la Belle Epoque*. Vol. 3. *Histoire de l'édition française* (Paris: Promodis, 1985), pp. 188–89.

Ambroise-Rendu, Anne-Claude, 'La Perception de la puissance française en 1900: l'exemple de l'Exposition universelle dans la presse', in Pierre Milza and Raymond Poidevin (eds), *La Puissance française à la 'Belle Epoque'. Mythe ou réalité?* (Paris: Éditions Complexe, 1992), pp. 143–60.

Anceau, Eric, *Napoléon III* (Paris: Éditions Tallandier, 2008).

Andrews, Howard, 'The Early Life of Paul Vidal de la Blache and the Makings of Modern Geography', *Transactions of the Institute of British Geographers* 11.2 (1986), pp. 174–82.

Ansay, Pierre, and René Schoonbrodt (eds), *Penser la ville. Choix de textes philosophiques* (Brussels: Archives d'architecture moderne, 1989).

Ariès, Philippe, 'L'Histoire des mentalités', in *La Nouvelle Histoire* (Paris: Éditions Complexe, 1988), pp. 167–90.

Artz, Frederick, *The Development of Technical Education* (Cleveland, OH: Society for the History of Technology, 1966).

Barbier, Frédéric, 'Les Imprimeurs', in Roger Chartier (ed.), *Le Temps des éditeurs du Romantisme à la Belle Epoque, Vol. 3: Histoire de l'édition française* (Paris: Promodis, 1985) pp. 68–89.

——, 'Libraires et colporteurs', in Roger Chartier (ed.) *Le Temps des éditeurs du Romantisme à la Belle Epoque Vol. 3: Histoire de l'édition française* (Paris: Promodis, 1985), pp. 229–59.

Barthes, Roland, *Mythologies* (Paris: Éditions de Seuil, 1957).

——, 'Rhétorique de l'image', *Communications*, 4 (1964), pp. 40–51.

——, 'Semiologie et urbanisme,' *Œuvres Complètes*, II (Paris: Seuil, 1967).

Bastié, Jean, *La Croissance de la banlieue parisienne* (Paris: PUF, 1964).

——, *Histoire de l'urbanisme à Paris* (Paris: Hachette, 1993).

——, *Nouvelle Histoire de Paris de 1914 à 1940* (Paris: Bibliothèque Historique de la Ville de Paris, 1997).

Belhoste, Bruno, Francine Masson, and Antoine Picon, *Le Paris des polytechniciens* (Paris: Délégation à l'action artistique de la ville de Paris, 1994).

Bellanger, Emmanuel, Annie Fourcaut, and Mathieu Flonneau (eds), '1919. Paris s'agrandir encore: annexion de la zone militaire et déclassement de l'enceinte fortifiée', *Paris/banlieue conflits et solidarités: historiographie, anthologie, chronologie 1788–2006* (Paris: Creaphis, 2007), pp. 160–64.

Benoît-Lévy, Georges, 'Plan d'extension de Paris', *La République française*, 19 juillet 1907.

Benjamin, Walter, *The Arcades Project*, trans. Howard Eiland and Kevin McLaughlin (Cambridge, MA: Belknap Press of Harvard University, 1999).

——, *Illuminations*, trans. Harry Zohn, ed. Hannah Arendt (New York: Schocken Books, 1969).

Berdoulay, Vincent, *La Formation de l'école française de géographie. Institut d'histoire, de géographie, d'économie urbaine de la ville de Paris* (Paris: Bibliothèque nationale, 1981).

Berthaut, Henri-Marie-Aguste, *La Carte de France* (2 vols) (Paris: Service géographique de l'armée, 1898).

Bertho-Lavenir, Catherine, 'L'Invention de la Bretagne: genèse d'un stéréotype', *ARSS*, 35 (1980), pp. 45–62.

——, *La Roue et le stylo: comment nous sommes devenus touristes* (Paris: Éditions Odile Jacob, 1999).

Bertin, Jacques, *Semiology of Graphics: Diagrams, Networks, Maps* (Madison, WI: University of Wisconsin Press, 1983).

Bigourdian, Guillaume, 'La Carte de France; d'après l'ouvrage de colonel Berthaut', *Annales de Géographie*, 8.42 (1899), pp. 427–37.

Bisson, G., *Mémoire sur les lotissements dans la région parisienne* (Paris: Institut d'urbanisme de l'Université de Paris, 1929).

Black, Jeremy, *Maps and History* (New Haven, CT and London: Yale University Press, 2000).

Blanc-Chaléard, Marie-Claude, 'L'Image de la puissance française dans les manuels d'histoire et de géographie autour de 1900', in Pierre Milza and Raymond Poidevin (eds), *La puissance française à la 'Belle Epoque'. Mythe ou réalité?* (Paris: Éditions Complexe, 1992), pp. 63–79.

Bonaparte, Louis-Napoléon (Empereur des Français), *Des idées napoléoniennes*, 6th ed. (London: W. Jeffs, 1860).

Bonnefont, Jean-Claude, 'La Lorraine dans l'œuvre de Paul Vidal de la Blache', in Paul Claval (ed.), *Autour de Vidal de la Blache. La formation de l'École française de géographie* (Paris: CNRS Éditions, 1993), pp. 81–88.

Bonnier, Louis, 'L'esthétique et l'architecture dans le concours pour le Plan d'aménagement et d'extension de Paris', *La Vie urbaine*, 4 (1920), pp. 29–38.

Booth, William, *In Darkest England and the Way Out* (London: Salvation Army, 1890).

Brantz, Dorothee, 'Recollecting the Slaughterhouse', *Cabinet*, 4 (2001), http://www.cabinetmagazine.org/issues/4/slaughterhouse.php.

Broc, Numa, 'L'Établissement de la géographie en France: diffusion, institutions, projets (1870–1890)', *Annales de Géographie*, 83 (1974), pp. 545–68.

——, 'Quelques débats dans la géographie française avant Vidal de la Blache', in Paul Claval (ed.), *Autour de Vidal de la Blache. La formation de l'École française de géographie* (Paris: CNRS Éditions, 1993), pp. 37–42.

Brousson, Jean-Jacques, 'La Mort des "fortifs"', *Cahiers de la République des lettres, des sciences et des arts, XII: Vers un nouveau Paris* (1930), p. 96.

Brown, Lloyd, *The Story of Maps* (Boston: Little, Brown, 1949).

Brunet, Roger, *La Carte mode d'emploi* (Paris: Fayard, 1987).

Brunot, André, *Le Corps des ponts et chaussées* (Paris: Éditions du Centre national de la recherché scientifique, 1982).

Buisson, F, *Dictionnaire de pédagogie et d'instruction primaire* (Paris: Hachette, 1882–1887).

Burch, R. M., and William Gamble, *Colour Printing and Colour Printers* (London: Pittman, 1910).

Burton, Richard D. E., *Blood in the City, Violence and Revolution in Paris, 1789–1945* (Ithaca, NY: Cornell University Press, 2001).

Butlin, R. A., *Geographies of Empire: European Empires and Colonies, 1880–1960* (Cambridge and New York: Cambridge University Press, 2009).

Calinescu, Matei, *Five Faces of Modernity* (Durham, NC: Duke University Press, 1987).

Cannon, James, *The Paris Zone: A Cultural History, 1840–1944* (Farnham: Ashgate Publishing, 2015).

Carmona, Michel, *Haussmann: His Life and Times, and the Making of Modern Paris* (Chicago, IL: Ivan R. Dee, 2002).

Caron, François, *Histoire des chemins de fer en France* (Paris: Fayard, 1997).

Caron, Remi, 'La Légende', *Cartes et figures de la terre* (Paris: Centre Georges Pompidou, Centre de Création Industrielle, 1980), pp. 329–31.

——, 'Les Choix du cartographe', *Cartes et figures de la terre* (Paris: Centre Georges Pompidou, Centre de Création Industrielle, 1980), pp. 9–15.

Casselle, Pierre, 'Le Régime législatif', in Roger Chartier (ed.), *Le Temps des éditeurs du Romantisme à la Belle Epoque, Vol. 3: Histoire de l'édition française* (Paris: Promodis, 1985), pp. 47–55.

Cate, Phillip Denis, *The Color Revolution: Color Lithography in France, 1890–1900* (Santa Barbara, CA: P. Smith, 1978).

——, *Prints Abound: Paris in the 1890s* (London: Lund Humphries Publishers, 2000).

Chabaud, Gilles (ed.), *Des guides imprimés du XVIe au XXe siècles* (Paris: Belin, 2000).

Champeaux, Antoine, *Michelin et l'aviation. Patriotisme industriel et innovation* (Panazol: Lavauzelle, 2006).

Charle, Christophe, *Paris fin de siècle* (Paris: Éditions de Seuil, 1998).

Chardon, Henri, *Communication à l'Académie des Sciences Morales et Politiques le 2 juillet 1927*, n.p.

Chartier, Roger, *Cultural History: Between Practice and Representation*, trans. Lydia G. Cochrane (Ithaca, NY: Cornell University Press, 1988).

——, 'Les Deux Frances: histoire d'une géographie', *Cahiers d'histoire*, 23 (1978), pp. 393–415.

——, 'Le monde comme représentation', *Annales, Économies, Sociétés, Civilisations*, 44.6 (1989), pp. 1505–20.

——, 'Text, Symbols and Frenchness', *Journal of Modern History*, 57 (1985), pp. 682–95.

Charvet, Marie, *Les Fortifications de Paris, 1880–1919* (Rennes: Presses universitaires de Rennes, 2005).

Chaudun, Nicholas, *Haussmann, Georges Eugène, préfet-baron de la Seine* (Arles: Actes Sud, 2009).

Chevalier, Louis, *Laboring Classes and Dangerous Classes in Paris during the First Half of the Nineteenth Century*, trans. Frank Jellinek (New York: Howard Fertig, 1973).

Chevallier, Fabienne, *Le Paris moderne: histoire des politiques d'hygiène* (Rennes: Presses universitaires de Rennes; Paris: Comité d'histoire de la ville de Paris, 2010).

Choay, Françoise, *The Modern City: Planning in the Nineteenth Century* (New York: G. Braziller, 1970).

Clark, T. J., 'The View from Notre Dame', *The Painting of Modern Life* (Princeton, NJ: Princeton University Press, 1984), pp. 23–70.

Claval, Paul, *Histoire de la géographie française de 1870 à nos jours* (Paris: Éditions Nathan, 1998).

Clerget, Pierre, 'L'Urbanisme: l'étude historique, géographique et économique', *Bulletin de la Société Neuchâteloise de Géographie*, XX (1913), pp. 213–31.

Cohen, Jean-Louis, and André Lortie, *Des fortifs au périf* (Paris: Picard/Pavillon de l'Arsenal, 1991).

Collet, M.J., '*Review: La Carte de France de l'Etat-Major (1887)*', *National Geographic Magazine*, 3 (1891), pp. 250–52.

Combreau, Yvan, *Paris et les élections municipales sous la Troisième République* (Paris: L'Harmattan, 1998).

Compte rendu sténographique de la séance inaugurale du Comité Supérieur de l'Aménagement et de l'Organisation générale de la région Parisienne, le mardi 19 juin 1928, au ministère de l'Intérieur, sous la présidence de M. Albert Sarraut. n.p.

Conklin, Alice L., *A Mission to Civilize: The Republican Idea of Empire in France and West Africa, 1895–1930* (Palo Alto, CA: Stanford University Press, 1997).

Conley, Tom, *The Self-Made Map: Cartographic Writing in Early Modern France* (Minneapolis, MN: University of Minnesota Press, 1996).

Conseil général du département de la Seine, *Procès-verbaux des délibérations* (Paris: Imprimerie municipale, 1893).

Cook, Karen Severud, 'From False Starts to Firm Beginnings: Early Color Printing of Geological Maps', *Imago Mundi*, 47 (1995), pp. 155–72.

Cook, Thomas, *Traveller's Handbook to Algeria and Tunisia* (London: Thomas Cook, 1926).

Cosgrove, Denis, *Mappings* (London: Reaktion Books, 1999).

Darnton, Robert, 'First Steps Towards a History of Reading', in *The Kiss of the Lamourette: Reflections in Cultural History* (New York: Norton, 1990), pp. 154–87.

——, *The Great Cat Massacre and Other Episodes in French Cultural History* (New York: Vintage Books, 1984).

——, 'Kiss of Lamourette', in *The Kiss of Lamourette: Reflections in Cultural History* (New York: Norton, 1990), pp. 3–20.

——, *Revolution in Print: The Press in France, 1775–1800* (Berkeley, CA: University of California Press, 1989).

——, 'What is the History of Books?', in *The Kiss of the Lamourette: Reflections in Cultural History* (New York: Norton, 1990), pp. 107–35.

Daubanton, Louis, 'Rapport relatif aux entreprises de construction à Paris et à l'interruption des travaux depuis cette année', in *Mémoires statistiques de la ville de Paris* (Paris: Imprimerie municipale, 1829).

Dausset, Louis, 'A propos de l'aménagement de la Région parisienne,' *Urbanisme*, special number (March 1932), pp. XIX–XX.

——, *Résumé du rapport au nom de la Première Commission et de la Commission mixte des fortifications, sur le déclassement total de l'enceinte fortifiée, l'annexion de la zone militaire, et sur le projet* (Paris: Imprimerie municipale, 1912), pp. 335–40.

De Certeau, Michel, *L'Invention du quotidien*, 2 vols (Paris: Union général d'éditions, 1980).

De Dainville, François, 'Les Bases d'une cartographie industrielle de l'Europe au XIX', in *L'Industrialisation en Europe au XIXe siècle, cartographie et typologie* (Paris: Éditions du Centre national de la recherche scientifique, 1972).

——, *Langage des géographes* (Paris: Éditions A. J. Picard et Cie, 1964).

De Margerie, Emm, 'La Nouvelle carte de France au 50000e: du Service géographique de l'Armée', *Annales de géographie*, 14.75 (15 May 1905), pp. 236–44.

De Margerie, Emm, and Louis Raveneau, 'La Cartographie à l'Exposition universelle de 1900', *Annales de Géographie*, 9.46 (15 July 1900), pp. 291–312.

De Moncan, Patrice, and Claude Heurteux, *Le Paris d'Haussmann* (Paris: Les Éditions du Mécène, 2002).

Dell, Simon, 'The Consumer and the Making of the Exposition Internationale des Arts Décoratifs et Industriels Modernes, 1907–1925', *Journal of Design History*, 12.4 (1999), pp. 311–25.

Delorme, Jean-Claude, 'Les Plans d'aménagement et d'extension des villes françaises', in *Les Cahiers de la recherche architecturale*, 8 (1981), pp. 10–29.

Demangeon, Albert, *Paris la ville et sa Banlieue* (Paris: Éditions Bourrelier et Cie, 1934).

Demarco, Eileen S, *Reading and Riding: Hachette's Railroad Bookstore Network in Nineteenth-Century France* (Bethlehem, PA: Lehigh University Press, 2006).

Demory, Hubert, 'Le Général Uhrich', *Le Village*, 2002, http://mapage.noos.fr/hubert.demory/uhrich.htm.

Dennis, Richard, *Cities in Modernity: Representations and Productions of Metropolitan Space, 1840–1930* (Cambridge: Cambridge University Press, 2008).

Desabres, Pascal, 'Prolonger le métro en banlieue: débats, enjeux politiques et réalisations, 1919–1946', in *Agrandir Paris, 1860–1970* (Paris: Publications de la Sorbonne, Comité d'histoire de la ville de Paris, 2012).

Desportes, Marc, *Paysages en mouvement* (Paris: Gallimard, 2005).

Desportes, Marc and Antoine Picon, *De l'espace au territoire, l'aménagement en France, 16e au 20e* (Paris: Presses de l'École nationale des ponts et chaussées, 1997).

Donald, James, *Imagining the Modern City* (Minneapolis, MN: University of Minnesota Press, 1999).

Driver, Felix, *Geography Militant: Cultures of Exploration and Empire* (Oxford and Malden, MA: Blackwell Publishers, 2001).

Duby, Georges, *Histoire de la France* (Paris: Larousse, 1995).

Duby, Georges, and Robert Mandrou, *Histoire de la civilisation française, XVIIe–XXe siècle*, Vol. 2 (Paris: Armand Colin, 1984).

Dumont, Marie-Jeanne, 'Du vieux Paris au Paris nouveau. Abécédaire d'architecture urbaine', *Le Débat*, 10 (1994), pp. 5–35.

Duncan, James S, 'Sites of Representation: Places, Time, and the Discourse of the Other', in James Duncand and David Ley (eds), *Place/Culture/Representation* (London: Routledge, 1993), pp. 39–56.

——, 'The Social Construction of Unreality: An Interactionist Approach to the Tourist's Cognition of Environment', in David Ley and Marwyn S. Samuels (eds), *Humanistic Geography* (Chicago, IL: Maaroufa Press, 1978), pp. 269–82.

Dunlop, Catherine, *Cartophilia* (Chicago, IL: University of Chicago Press, 2014).

Dupuy, Aimé, 'Les Livres de lecture de G. Bruno', *Revue d'histoire économique et sociale*, 31.2 (1953), pp. 128–51.

Eco, Umberto, 'Function and Sign: The Semiotics of Architecture', in *Signs, Symbols, and Architecture* (New York: Wiley, 1980), pp. 11–69.

Edney, 'Brian Harley's Career and Intellectual Legacy', *Cartographica*, 40.1–2 (2005), pp. 2–98.

——. 'Theory and the History of Cartography', *Imago Mundi*, 48 (1996), pp. 185–91.

Ehrensvärd, Ulla, 'Color in Cartography: A Historical Survey', in David Woodward (ed.), *Art and Cartography, Six Historical Essays* (Chicago, IL: University of Chicago Press, 1987), pp. 123–46.

Elliot, James, *The City in Maps: Urban Mapping to 1900* (London: The British Library, 1987).

Evenson, Norma, *Paris: A Century of Change, 1878–1978* (New Haven: Yale UP, 1979).

Ferguson, Priscilla Parkhurst, 'Reading City Streets', *The French Review*, 61.3 (1988), pp. 386–97.

Ferry, Jules, *Les Comptes fantastiques d'Haussmann* (Paris and Neuilly-sur-Seine: G. Durier, 1868).

Fierro, Alfred, 'Les Guides de voyages au XIXe siècle', in Roger Chartier (ed.), *Histoire de l'édition française* (Paris: Promodis, 1983), p. 193.

——, *Histoire et dictionnaire de Paris* (Paris: Éditions Robert Laffont, 1996).

Foncin, Myriem, 'Quelques réflexions géographiques à propos du concours (1re section) pour le Plan d'aménagement d'extension de Paris', *La Vie urbaine*, 5 (1920), pp. 77–89.

Foucault, Michel, *Discipline and Punish: The Birth of the Prison* (Harmondsworth: Penguin, 1991).

——, *L'Ordre du discours* (Paris: Gallimard, 1971).

——, 'Questions on Geography', in *Power/Knowledge* (1980), pp. 63–78.

——, 'Space, Knowlege, and Power', in *The Foucault Reader*, ed. Paul Rainbow (New York: Pantheon, 1984).

Fouillée, Augustine (G. Bruno), *Le Tour de la France par deux enfants* (Paris: Eugene Belin Libraire Classique, 1905).

Fourcaut, Annie et al., *Paris/banlieue: conflits et solidarités: historiographie, anthologie, chronologie, 1788–2006* (Paris: Creaphis, 2007).

Francon, Marc, *Le Guide Vert Michelin. L'invention du tourisme culturel populaire* (Paris: Economica, 2001).

Friedman, Susan W., 'Marc Bloch, Sociology and Geography', *Cambridge Studies in Historical Geography*, Vol. 24 (Cambridge and New York: Cambridge University Press, 1996).

Furet, François, *Penser la Révolution française* (Paris: Gallimard, 1978).

Gade, Daniel W., 'Gallic Trajectories of Geography and the Self', *The Geographical Review*, 89.1 (1999), pp. 122–28.

Gaudin, Jean-Pierre, *L'Avenir en plan technique et politique dans la prévision urbaine, 1900–1930* (Paris: Éditions Champ Vallon, 1985).

Gibel, Pierre, 'La Région parisienne', *La Vie urbaine*, 55 (1950), pp. 115–54.

Gibrat, Joseph, *Traité de la géographie moderne* (Avignon: Chaillot Etienne, 1813).

Giddens, Anthony, and Christopher Pierson, *Conversations with Anthony Giddens, Making Sense of Modernity* (Stanford, CA: Stanford University Press, 1998).

Girardet, Raoul, *Le Nationalisme français* (Paris: Éditions du Seuil, 1983).

Godlewska, Anne, *Geography Unbound: French Geographic Science from Cassini to Humboldt* (Chicago, IL: University of Chicago Press, 1999).

Green, Nicholas, *The Spectacle of Nature: Landscape and Bourgeois Culture in the Nineteenth Century* (New York: St. Martin's Press, 1990).

Gritti, Jules, 'Les Contenus culturels du guide bleu', *Communications*, special number 'Vacances et tourismes' (1967), pp. 51–64.

Guérard, Albert, *L'Avenir de Paris* (Paris: Payot, 1929).

Guiomar, Jean-Yves, 'Le Tableau de la géographie de la France de Vidal de la Blache', in Pierre Nora (ed.), *Les Lieux de mémoire, II. La Nation* (Paris: Gallimard, 1986), pp. 568–97.

Hall, Sir Peter, *Cities in Civilization* (New York: Routledge, 1998).

Hancock, Claire, *Paris et Londres au XIXe siècle, représentations dans les guides et récits de voyage* (Paris: CNRS Éditions, 2003).
Harley, J. B., *The New Nature of Maps, Essays in the History of Cartography* (Baltimore, MD: The Johns Hopkins University Press, 2001).
Harvey, David, *Paris: Capital of Modernity* (New York: Routledge, 2003).
——, *The Urban Experience* (Baltimore, MD: The Johns Hopkins University Press, 1989).
Harvey, P. D. A., *The History of Topographical Maps* (London: Thames and Hudson, 1980).
Haussmann, Baron Georges Eugène, *Mémoires, Édition intégrale* (Paris: Seuil, 2000).
Hazan, Eric, *L'Invention de Paris: il n'y a pas de pas perdus* (Paris: Seuil, 2002).
Helgerson, Richard, 'The Land Speaks: Cartography, Chorography and Subversion in Renaissance England', *Representations*, 16 (1986), pp. 51–85.
Herbert, Robert L., *Impressionism: Art, Leisure, and Parisian Society* (New Haven, CT: Yale University Press, 1988).
Hénard, Eugène, 'L'Embellissement des villes', *L'Architecture* (7 May 1910), pp. 341–43.
——, 'Les Espaces libres et les fortifications', *L'Architecture* (December 1909), pp. 333–39.
——, *Les Etudes sur les transformations de Paris* (Paris: Éditions l'Equerre, 1982).
Hennequin, Frédéric, *La Topographie mise à la portée de tous* (Paris: Librairie militaire de J. Dumaine, 1875).
Higgonet, Patrice, *Paris: Capital of the World* (Cambridge, MA: Harvard University Press, 2002).
Hobsbawm, Eric, *The Invention of Tradition* (Cambridge: Cambridge University Press, 1983).
——, *Nations and Nationalism Since 1780* (Cambridge: Cambridge University Press, 1990).
Hooper, Barbara, '"Split at the Roots": A Critique of the Philosophical and Political Sources of Modern Planning Doctrine', *Frontiers: A Journal of Women Studies*, 13.1 (1992), pp. 45–80.
Horne, Alistair, *Seven Ages of Paris* (New York: Knopf, 2002).
Horne, Janet, *A Social Laboratory for Modern France: The Musée Social* (Durham, NC: Duke University Press, 2002).
Jacob, Christian, *L'Empire des cartes: approche théorique de la cartographie à travers l'histoire* (Paris: Éditions Albin Michel, 1992).
——, *The Sovereign Map: Theoretical Approaches in Cartography throughout History* (Chicago, IL: University of Chicago Press, 2006).
——, 'Toward a Cultural History of Cartography', *Imago Mundi*, 48 (1996), pp. 191–98.
Jakobson, Roman, *Essais de linguistique générale* (Paris: Seuil, 1963).
Jaussely, Léon. 'Chronique de l'Urbanisme', *La Vie urbaine* (1919), pp. 181–202.
Joanne, Adolphe, *Paris Illustré* (Paris: Hachette, 1870).
Joanne, Paul, *Alger et ses environs* (Paris: Hachette, 1908).
——, *Géographie de l'Algérie* (Paris: Hachette, 1908).

Joly, Martine, *Introduction à l'analyse de l'image* (Paris: Nathan, 1993).

Jonas, Raymond. A., 'Monument as Ex-Voto, Monument as Historiography: The Basilica of Sacré-Cœur', *French Historical Studies*, 18.2 (1993), pp. 482–502.

Jordan, David, *Transforming Paris: The Life and Labors of Baron Haussmann* (Chicago, IL: University of Chicago Press, 1995).

Joyce, Patrick, *The Rule of Freedom: Liberalism and the Modern City* (London: Verso, 2003).

King, Geoffrey, *Mapping Reality: An Exploration of Cultural Cartographies* (New York: St. Martin's Press, 1996).

Konvitz, Josef W., *Cartography in France, 1660–1848: Science, Engineering, and Statecraft* (Chicago, IL: University of Chicago Press, 1987).

——, 'Remplir la carte', trans. Paule Guivarch, in *Cartes et figures de la terre* (Paris: Centre Georges Pompidou, 1980), pp. 304–14.

Koshar, Rudy (ed.), *Histories of Leisure* (Oxford: Berg, 2002).

Kunstler, James Howard, *The City in Mind* (New York: The Free Press, 2001).

Laboulais, Isabelle (ed.), *Les Usages des cartes (XVII–XIXe): pour une approche pragmatique des productions cartographiques* (Strasbourg: Presses universitaires de Strasbourg, 2008).

Lacaze, Jean-Paul, *Paris, urbanisme d'état et destin d'une ville* (Paris: Flammarion, 1994).

Laisney, François, 'Réglementer la banlieue? L'exemple du plan Prost', *Les Cahiers de la recherche architecturales*, 38–39 (1996), pp. 117–30.

Lambert, Michèle, 'Le Chemin de fer et la banlieue (1837–1914)', *Cahiers de la recherché architecturale*, 38–39 (1996), pp. 95–106.

Landau, Bernard, and Annie Térade, 'La Fabrication des rues de Paris', in Bertrand Lemoine and Marc Mimram (eds), *Paris d'ingénieurs* (Paris: Pavillon de l'arsenal, 2000), pp. 91–105.

Latour, François, *L'Aménagement du Grand Paris, où en sommes-nous? Rapport au conseil municipal* (Paris: Imprimerie municipale, 1934).

Lavedan, Pierre, *Géographie des villes* (Paris: Gallimard, 1959).

——, *L'Histoire de Paris, Que sais-je?* (Paris: Presses Universitaires de Paris, 1967).

——, *Nouvelle histoire de Paris: histoire de l'urbanisme à Paris* (Paris: Hachette, 1993).

——, *Les Villes françaises* (Paris: Éditions Vincent Fréal, 1960).

Le Goff, Jacques, 'L'Histoire nouvelle', in *La Nouvelle Histoire* (Paris: Éditions Complexe, 1988), pp. 35–73.

LeBras, Hervé, *Les Trois France* (Paris: Odile Jacob, 1995).

Lefebvre, Henri, *The Production of Space* (Oxford: Blackwell, 1991).

Lehman, Nicholas, 'Atlas Shrugs', *The New Yorker*, 9 April 2001, pp. 131–34.

Lehning, James R., *To Be a Citizen: The Political Culture of the Early French Third Republic* (Ithaca, NY: Cornell University Press, 2001).

Lejeune, Dominique, *La France de la Belle Epoque, 1896–1914* (Paris: Armand Colin, 1995).

Lemoine, Bertrand, and Marc Mimram (eds), *Paris d'ingénieurs* (Paris: Pavillon de l'arsenal, 2000).

Lentz, Thierry, *Napoléon III, Que sais-je?* (Paris: PUF, 1995).
Lestringant, Frank, 'Suivez la guide', in *Cartes et figures de la terre* (Paris: Centre Georges Pompidou, Centre de Création Industrielle, 1980), pp. 424–35.
Levin, Miriam, 'Inventing Modern Paris: The Dynamic Relationship between Expositions, Urban Development and Museums', *Quaderns d'Historia de l'Enginyeria*, 13 (2012), pp. 35–56.
Lojkine, Jean, *La Politique urbaine dans la région parisienne, 1945–1971* (Paris: Mouton, 1972).
Lyons, Martyn, *Le Triomphe du livre: une historie sociologique de la lecture dans la France du XIXe siècle* (Paris: Promodis, 1987).
Mabilleau, Gilles, 'Section d'hygiène urbaine et rurale. Séance constitutive du 14 janvier 1908', *Annales du Musée social* (1908), p. 56.
MacCannell, Dean, *The Tourist: A New Theory of the Leisure Class* (Berkeley, CA: University of California Press, 2001).
Mansel, Philip, *Paris between Empires, 1814–1852* (London: John Murray, 2001).
Marchand, Bernard, *Les Ennemis de Paris: la haine de la grande ville des Lumières à nos jours* (Rennes: Presses universitaires de Rennes, 2009).
——, *Paris, histoire d'une ville (XIXe–XXe siècle)* (Paris: Seuil, 1993).
Marin, Louis, *De la représentation* (Paris: Gallimard, 1994).
Martin, Geoffrey J., and Preston E. James, *All Possible Worlds: A History of Geographical Ideas* (New York: John Wiley and Sons, 1993).
Martin, Henri Jean, and Roger Chartier (eds), *Histoire de l'édition française*. 4 vols. (Paris: Promodis, 1986).
Martin, Odile, and Henri Jean, 'Le Monde des éditeurs', in Henri Jean and Roger Chartier (eds), *Histoire de l'édition française, Vol. 3: Le temps des éditeurs* (Paris: Promodis, 1983), pp. 176–244.
Meissas, Achille and Aug. Michelot, *Nouvelle géographie méthodique* (Paris: Hachette, 1854).
Mélonio, Françoise, *Naissance et affirmation d'une culture nationale* (Paris: Éditions du Seuil, 1998).
Melot, Michel, 'Le Texte et l'image', in Roger Chartier (ed.), *Le Temps des éditeurs du Romantisme à la Belle Epoque, Vol. 3: Histoire de l'édition française* (Paris: Promodis, 1985), pp. 286–311.
Mettie, Erik, *World's Fairs* (Princeton, NJ: Princeton Architectural Press, 1998).
Meynier, André, *Histoire de la pensée géographique en France (1872–1969)* (Paris: Presses Universitaires de France, 1969).
Michelin, *Le Guide Michelin* (Clermont Ferrand: Michelin, 1900).
Milo, Daniel, 'Les noms des rues', in Pierre Nora (ed.), *Les Lieux de mémoire, II. La Nation* (Paris: Gallimard, 1986), pp. 283–315.
Mollier, Jean Yves, *Louis Hachette, 1800–1864: le fondateur d'un empire* (Paris: Fayard, 1999).
Monmonier, Mark, *From Squaw Tit to Whorehouse Meadow: How Maps Name, Claim, and Inflame* (Chicago, IL: University of Chicago Press, 2006).
——, *How to Lie with Maps*, 2nd ed. (Chicago, IL and London: University of Chicago Press, 1996).

Monsarrat, M. G., 'La Commission Supérieure d'aménagement, d'embellissement et d'extension des villes au Ministère de l'Intérieur', *Urbanisme*, special number (March 1932), pp. xiii–xviii.

Morizet, André, 'Comment aménager une ville', *L'Architecture aujourd'hui* (June 1937), p. 97.

Morton, Patrcia A., 'National and Colonial: The Musée des Colonies at the Colonial Exposition, Paris, 1931', *Art Bulletin*, 80.2 (1998), pp. 357–77.

Nebenzahl, Kenneth, *The Imperial Map: Cartography and Masters of Empire* (Chicago, IL: University of Chicago Press, 2009).

Nora, Pierre, 'Entre mémoire et histoire. La problématique des lieux', in *Les Lieux de mémoire, I. La République* (Paris: Gallimard, 1984), pp. xvii–xlii.

Nordman, Daniel, 'Les Guides-Joanne, ancêtres des Guides Bleus', in Pierre Nora (ed.), *Les Lieux de Mémoire, II. La Nation* (Paris: Gallimard, 1986), pp. 1035–72.

Ogborn, Miles, *Spaces of Modernity* (New York and London: The Guilford Press, 1998).

Offerlé, Michel, 'Illégitime et légitimation du personnel politique ouvrier en France avant 1914', *Annales. Économies, Société, Civilisations*, 39.4 (1984), pp. 681–716.

Olson, Kory, 'A New Paris? Eugène Hénard's Vision of Paris's Future', *Contemporary French and Francophone Studies*, 14.4, special issue 'Seismes/ Seismic Shifts' (September 2010), pp. 431–40.

——, 'Creating Map Readers: Republican Geographic and Cartographic Discourse in G. Bruno's (Augustine Fouillée) 1905 *Le Tour de la France par deux enfants*', *Modern and Contemporary France*, 19.1 (February 2011), pp. 37–51.

——, 'Driving a New Tourist: Reading Maps in early Guides Michelin France', *Imago Mundi: The International Journal for the History of Cartography*, 62.2 (June 2010), pp. 205–20.

Ormeling, Ferjan, 'Toponymies', trans, Services linguistiques du ministère des Affaires étrangères à La Haye, in *Cartes et figures de la terre* (Paris: Centre Georges Pompidou, Centre de Création Industrielle, 1980), pp. 332–34.

Ory, Pascal, 'Le Centenaire de la Révolution Française', in Pierre Nora (ed.), *Les Lieux de Mémoire, II. La Nation* (Paris: Gallimard, 1986), pp. 523–60.

——, *L'Expo Universelle, 1889* (Paris: Éditions Complexe, 1989).

——, *Les Expositions universelles de Paris* (Paris: Éditions Ramsay, 1982).

Ozouf, Jacques, and Mona Ozouf, '*Le Tour de la France par deux enfants*: le petit livre rouge de la République', in Pierre Nora (ed.), *Les Lieux de mémoire, I. La République* (Paris: Gallimard, 1984), pp. 291–321.

Ozouf, Mona, *L'Ecole, l'Eglise et la République, 1871–1914* (Paris: Éditions Cana/ Jean Offredo, 1982).

Papayanis, Nicholas, *Planning Paris before Haussmann* (Baltimore, MD and London: The Johns Hopkins University Press, 2004).

Pearson, Karen S., 'The Nineteenth-Century Colour Revolution: Maps in Geographical Journals', *Imago Mundi*, 32 (1980), pp. 9–20.

Peirce, Charles S., *The Essential Peirce: Selected Philosophical Writings*, Vol. 2 (Bloomington, IN: Indiana University Press, 1992).

Pelletier, Monique, *Les Cartes de Cassini: la science au service de l'État et des régions* (Paris: Éditions CTHS, 2002).

Pelletier, Monique, and Henriette Ozanne, *Portraits de la France: les cartes témoins de l'histoire* (Paris: Hachette, 1995).

Perrot, Philippe, 'Du luxe au bien-être au XIXe siècle en France', *Ethnologie française*, 22.2 (April–June 1992), pp. 187–95.

Peters, Jeffrey N., *Mapping Discord: Allegorical Cartography in Early Modern French Writing* (Newark, DE: University of Delaware Press, 2004).

Petitfrere, Claude, 'Une ville mise en scène: Tours d'après l'iconographie générale des XVIe–XVIIIe siècle', in Claude Petitfrere (ed.), *Images et imaginaires de la ville* (Tours: Maison des Sciences de la Ville Université François-Rabelais, 1998), pp. 175–201.

Pickles, John, *A History of Spaces: Cartographic Reason, Mapping, and the Geo-Coded World* (London and New York: Routledge, 2004).

Picon, Antoine, 'Les Fortifications de Paris', in *Le Paris des Polytechniciens* (Paris: Délégation à l'Action artistique de la ville de Paris, 1994), pp. 213–21.

——, 'Nineteenth-Century Urban Cartography and the Scientific Ideal: The Case of Paris', *Osiris*, 2nd Series, 18 (2003), pp. 135–49.

——, *Le Paris des ingénieurs saint-simoniens, Paris d'ingénieurs* (Paris: Éditions du Pavillon de l'Arsenal, 1995), pp. 37–43.

Picon, Antoine, and Jean-Paul Robert, *Un atlas parisien, le dessus des cartes* (Paris: Éditions du Pavillon de l'Arsenal, 1999).

Pinon, Pierre, *Paris: biographie d'une capitale* (Paris: Éditions Hazan, 1999).

Pinon, Pierre, and Bertrand Le Boudec, *Les Plans de Paris, histoire d'une capitale* (Paris: BNF, 2004).

Poëte, Marcel, *Introduction à l'urbanisme: l'évolution des villes, la leçon de l'histoire, l'antiquité* (Paris: Boivin et Cie, 1929).

Price, Roger, *A Concise History of France* (Cambridge: Cambridge University Press, 1993).

Prochasson, Christophe, *Paris 1900, Essai d'histoire culturelle* (Paris: Calmann-Lévy, 1999).

'Le Programme et le résultat du concours pour le plan d'aménagement et d'extension de Paris', *La Vie urbaine* (1920), pp. 91–111.

Rabinow, Paul, *French Modern: Norms and Forms of the Social Environment* (Cambridge, MA: MIT Press, 1989).

Rearick, Charles, 'Public Roads of Light, Water, and Drains', in *Pleasures of the Belle Epoque: Entertainment and Festivity in Turn-of-the-Century France* (New Haven, CT: Yale University Press, 1985).

Remaury Pierre, and Jean Royer, 'Région Parisienne', *L'Œuvre de Henri Prost, Architecture et urbanisme* (Paris: Académie d'Architecture, 1960), pp. 143–82.

Renan, Ernest, *Qu'est-ce qu'une nation?* (Paris: Mille et une nuits, 1997).

Richez, Sébastien, 'Le Facteur rural des postes en France avant 1914: un nouveau médiateur au travail', *Le Mouvement social*, 218 (2007), pp. 29–44.

Roncayolo, Marcel, 'Le Département', in Pierre Nora (ed.), *Les Lieux de mémoire, II. La Nation* (Paris: Gallimard, 1997), pp. 2937–74.

Roncayolo, Marcel et al., *La Ville aujourd'hui: mutations urbaines, décentralisation et crise du citadin* (Paris: Seuil, 2001).

Rouleau, Bernard, *Paris: histoire d'un espace* (Paris: Seuil, 1997).

Rouvet, Jean-Baptiste Massillon, *Viollet-le-Duc et Alphand au siège de Paris* (Paris: Librairies Imprimeries Réunies, 1892), pp. 19–22.

Royer, Jean, 'Henri Prost urbaniste 1928 Région parisienne', *Urbanisme*, 88 (1965), pp. 21–24.

Saalman, Howard, *Haussmann: Paris Transformed* (New York: G. Braziller, 1971).

Saint-Simon, Claude Henri comte de, *Selected Writings on Science, Industry and Social Organisation*, ed. and trans. Keith Taylor (New York: Holmes and Meier Publishers, Inc, 1975).

Sawyer, Stepthen W., 'Locating the Centre: Confining and Defining the Power of Paris', *Nottingham French Studies*, 44 (2005), pp. 20–37.

Schwartz, Vanessa R., *Spectacular Realities: Early Mass Culture in Fin-de-Siècle Paris* (Berkeley, Los Angeles, and London: University of California Press, 1998).

Sellier, Henri, *Les Banlieues urbaines et la réorganisation administrative du département de la Seine* (Paris: Librairie des sciences politiques et sociales, Marcel Rivière et Cie, 1920).

Service géographique de l'armée, *Carte d'état-major type 1900. Les environs de Paris, dites Carte de France* (Paris: Service géographique de l'armée, 1906).

Siestrunck, René, 'La Carte militaire', in *Cartes et figures de la terre* (Paris: Centre Georges Pompidou, 1980), pp. 363–74.

——, 'Tourisme, patriotisme, et topographie', in *Cartes et figures de la terre* (Paris: Centre Georges Pompidou, 1980), pp. 346–47.

Sims, George R., *How the Poor Live* (London: Chatto & Windus, 1883).

Soulignac, Françoise, *La Banlieue parisienne* (Paris: La Documentation française, 1993).

Sowerwine, Charles, *France Since 1870: Culture, Politics and Society* (New York: Palgrave, 2001).

Strachan, John, 'Romance, Religion and the Republic: Bruno's *Le Tour de la France par deux enfants*', *French History*, 18.1 (2004), pp. 96–118.

Stovall, Tyler, *The Rise of the Paris Red Belt* (Berkeley, CA: University of California Press, 1990).

Sutcliffe, Anthony, *The Autumn of Central Paris: The Defeat of Town Planning, 1850–1970* (Montreal: McGill Queen's University Press, 1971).

——, *Towards a Planned City* (New York: St. Martin's Press, 1981).

Térade, Annie, and Michèle Lambert-Bresson (eds), *Villes françaises au XIXe siècle: Aménagement, extension et embellissement* (Paris: Éditions Recherches/ IPRAUS, coll. Les Cahiers de l'IPRAUS, 2002).

Tesnière, Valérie, 'L'Édition universitaire', in Roger Chartier (ed.), *Le Temps des éditeurs du Romantisme à la Belle Époque. Vol. 3: Histoire de l'édition française* (Paris: Promodis, 1985), pp. 217–27.

Thacker, Andrew, *Moving through Modernity: Space and Geography in Modernism* (Manchester and New York: Manchester University Press, 2003).

Thiesse, Anne-Marie, *La Création des identités nationales: Europe XVIIIe–XXe siècle* (Paris: Seuil, 1999).

Thomas, Albert, 'Préface', in Henri Sellier, *Les Banlieues urbaines et la réorganisation administrative du département de la Seine* (Paris: Librairie des sciences politiques et sociales, Marcel Rivière et Cie, 1920), pp. 5–11.

Thompson, Christopher S., *The Tour de France: A Cultural History* (Berkeley, CA: University of California Press, 2006).

Thrower, Norman, *Maps and Civilization* (Chicago, IL and London: University of Chicago Press, 1996).

Tooley, Ronald, *Maps and Map-makers* (New York: Dorset Press, 1987).

Topalov, Christian, 'The City as Terra Incognita: Charles Booth's Poverty Survey and the People of London, 1886–1891', *Planning Perspectives*, 8 (1993), pp. 395–425.

Turrel, Denise, 'La Couleur de la ville. Les représentations urbaines dans les cartes de l'ancien régime', in Claude Petitfrere (ed.), *Images et imaginaires de la ville* (Tours: Maisons des Sciences de la Ville Université François-Rabelais, 1998), pp. 125–39.

Unwin, Tim, *The Place of Geography* (New York: John Wiley and Sons, 1992).

Van Damme, Stéphane, *Métropoles de papier. Naissance de l'archéologie urbaine à Paris et à Londres (XVIIe–XXe siècles)* (Paris: Les Belles Lettres, 2012).

Vayassière, Bruno-Henri, '"La" Carte de France', in *Cartes et figures de la terre* (Paris: Centre Georges Pompidou, 1980), pp. 252–65.

Veblen, Thorstein, *The Theory of the Leisure Class* (New York: Dover Publications, 1994).

Verdet, Jean-Pierre, 'Vous êtes ici', in *Cartes et figures de la terre* (Paris: Centre Georges Pompidou, Centre de Création Industrielle, 1980), pp. 68–75.

Vidal de la Blache, Paul, 'La Carte de France au 50000[e]', *Annales de Géographie*, 68 (1904), pp. 113–20.

——, 'La Conception actuelle de l'enseignement de la géographie', *Annales de Géographie*, 75 (1905), pp. 193–207.

——, 'Des divisions fondamentales du sol français', *Bulletin littéraire*, 2:1 (1888), pp. 1–7, 49–57.

——, *France: départements* (Paris: Armand Colin, 1885).

——, *La France et les Cinq parties du monde (Les cartes murales)* (Paris: Armand Colin, 1885–1886).

——, 'Leçon d'ouverture du cours de géographie', *Annales de Géographie*, 38 (1899), pp. 97–109.

——, *Tableau de la géographie de la France* (Paris: La Table Ronde, 1904, 1994).

Vivier, Nadine, Noëlle Dauphin, Gilles Pécout, and Brigitte Waché, *Dictionnaire de la France du XIXe siècle* (Paris: Hachette, 2002).

Wakeman, Rosemary, *The Heroic City: Paris, 1945–1958* (Chicago, IL: University of Chicago Press, 2009).

Weber, Eugen, *France, Fin de Siècle* (Cambridge, MA: Harvard University Press, 1986).

——, 'L'Hexagone', in Pierre Nora (ed.), *Les Lieux de mémoire, I. La République* (Paris: Gallimard, 1984), pp. 1171–90.

——, *Peasants into Frenchmen: The Modernization of Rural France, 1870–1914* (Stanford, CA: Stanford University Press, 1976).

Weeks, Willet, *The Man Who Made Paris Paris, The Illustrated Biography of Georges Eugene Haussmann* (London: London House, 1999).

Welch, Edward, 'Apprehending the *banlieue*: *Les Passagers du Roissy-Express* and spatial enquiry in contemporary France', *Francosphères*, 3 (2014), pp. 173–86.

——, 'La Carte et le Territoire: Mapping, Photography and the Visualization of Contemporary France', *Nottingham French Studies*, 53.2 (2014), pp. 186–200.

Wieviorka, Olivier, and Christopher Prochasson, *La France du XXe siècle: documents d'histoire* (Paris: Éditions du Seuil, 1994).

Wilford, John Noble, *The Mapmakers* (New York: A. A. Knopf, 2000).

Wolf, Peter, *Eugène Hénard and the Beginning of Urbanism in Paris, 1900–1914* (The Hague: International Federation for Housing and Planning; Paris: Centre de recherché d'urbanisme, 1968).

Wood, Dennis, 'Introducing the Cartography of Reality', in David and Marywn S. Samuels Ley (eds), *Humanistic Geography, Prospects and Problems* (Chicago, IL: Maaroufa Press, 1978), pp. 207–19.

——, *The Power of Maps* (New York: The Guilford Press, 1992).

Zimmer, Oliver, *Nationalism in Europe, 1890–1940* (Basingstoke: Palgrave Macmillan, 2003).

Index